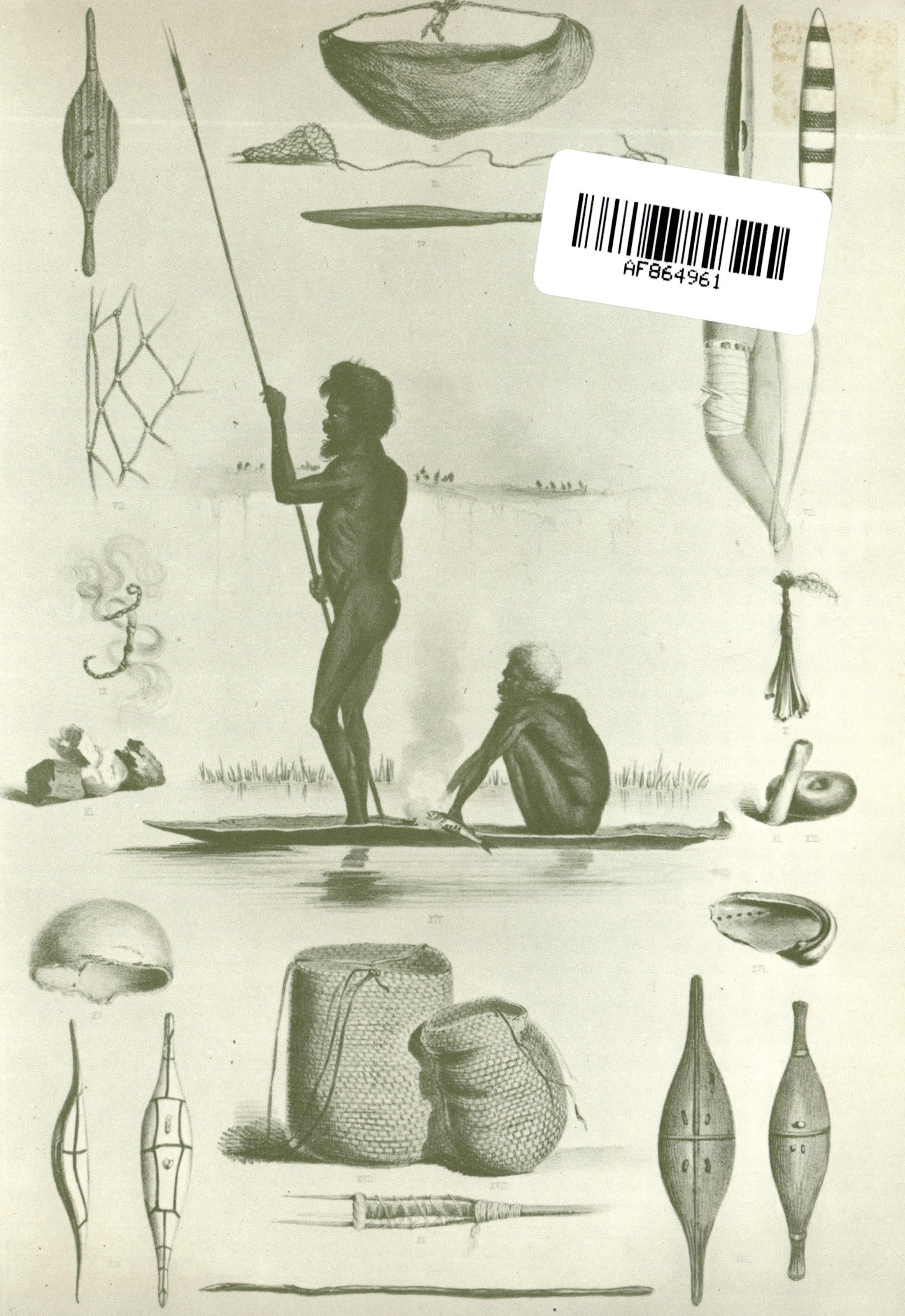

Aboriginal Bark Canoes
of the MURRAY VALLEY

To. Martha + John Collet.
With Compliments &
Kindest Regards
Alex [illegible]
Sydney.
5/5/76

This outstanding canoe tree on the Edson property at Blanchetown, South Australia, stands as a memorial to the Moorundie Aborigines

Aboriginal Bark Canoes

of the *MURRAY VALLEY*

Robert Edwards

Curator of Anthropology,
South Australian Museum

Robert Edwards

Published for the South Australian
Museum by

RIGBY

First published 1972 by
Rigby Limited • Adelaide • Sydney
Melbourne • Brisbane • Perth
for the
SOUTH AUSTRALIAN MUSEUM

Library of Congress Catalog Card Number 72-83023
National Library of Australia Registry
Card Number & ISBN 0 85179 412 2

Wholly designed and set up in Australia
Printed in Hong Kong

FOREWORD

This book describes the canoes of the Aborigines of Australia. In particular, it tells the story of the simple bark canoe of the Murray Valley and the large river population whose needs it served for many centuries. This apparently rudimentary craft they developed to extend their quest for food over the vast Murray-Darling basin.

The written history of the bark canoe is concurrent with the history of the European settlement of Australia and the early explorers who reported fleeting glimpses of them as Aborigines made their escape. The indigenous Australians had a great skill in manoeuvring their flimsy craft which astounded many early observers, and its practicability for efficient river travel was recognised by many inland explorers who made vain attempts to duplicate their construction. The flow of settlers along the inland watersheds hastened the disappearance of the Aborigines and their canoes. Ironically these same settlers, who lacked skill in the fabrication of bark canoes, were the ones who found them particularly useful for navigating their river frontages.

The advent of small boats followed by the massive paddle steamers spelt doom for the canoe and its maker.

By the turn of the century it was a rare sight to see an Aboriginal gliding along the river in his frail bark. Eventually the Aborigines themselves turned to European craft for transport.

Today all that remains are the large river gums with their scars. But these are important reminders that for thousands of years Australia was a land of Aborigines. Now in less than 200 years the indigenous culture has been all but destroyed. It is a responsibility and a privilege to make certain that the few remaining relics of this long occupation are preserved for the interest and enjoyment of future generations.

W. Grant Inglis
Director of Environment
and Conservation, South Australia

A number of huge scarred river gums, whether alive or dead, are the sole remnants of an important traditional industry of the hunter-gatherers who foraged the Australian landscape for many thousands of years before the arrival of Europeans. We call these gums "Canoe Trees." They are abundant along the great rivers and tributaries of south-eastern Australia. Some, centuries old, live on as noble memorials to a people who have all but disappeared. Others are the stark, bleached and decaying bones of great red gums drowned by the harnessing of the Murray River to the needs of modern society.

From these trees Australian Aborigines cut, prised and peeled the bark sheets that they made into their boats. White men laughed at these rude and simple craft, yet many lived to be thankful for them and to marvel at the skill with which the Aborigines applied their simple stone and wooden tools to their fabrication. No one knows for how many centuries they plied their skills, because a canoe's life was short and even the trees from which they came could not endure forever—especially after the arrival of Europeans. Trees that had the maturity of centuries disappeared rapidly before the onslaught of land clearance and the consuming fires of the paddle steamers.

For some years the South Australian Museum, assisted by the late Dr E. Couper Black, has photographed and catalogued canoe trees within the State and created a new awareness of their importance to our nation's story.

When Europeans arrived in Australia, watercraft were in use in nearly all parts of the continent where large bodies of water existed. When Abel Tasman visited the north-west in 1642, he reported them as crude and raft-like (Muller, 1898). In addition to a range of wood, reed and bark rafts which varied from region to region, later explorers noted a number of distinct types of canoes.

During their voyage of discovery in 1802 the French navigators Péron and Freycinet found the Aborigines at Port Jackson, New South Wales, making extensive use of bark canoes

From a lithograph by Charles Lesueur

The simplest craft consisted of a large sheet of thick bark stripped from a river red gum, *Eucalyptus camaldulensis*. The two ends were pointed and the bark manipulated, while still fresh and pliable, to form a very crude boat which was used for crossing rivers and for fishing. Usually it was propelled by punting with a long pole. The fragile nature of these simple craft limited their use to the protected inland waters of the Murray-Darling basin, western Victoria, and the south-east of South Australia. Immersion in calm waters for extended periods caused the bark sheets to become waterlogged.

On the rivers and creeks of coastal New South Wales and in south-eastern Victoria, explorers encountered a second type of canoe fabricated from large sheets of comparatively thin bark (McCarthy, 1944, pp. 184-187). To make these canoes, the Aborigines removed a cylinder of bark three metres to five metres long from a tall straight eucalypt, usually a stringybark, *Eucalyptus obliqua*. After the heavy sheet had been prised off, it was laid on logs. The rough

This sketch drawn by Captain Charles Sturt shows the junction of the Darling and Murray Rivers: the point where he named the Murray River in January, 1830

Tied bark canoes in use at Port Jackson, New South Wales, in 1802

outer surface was stripped and the ends thinned in preparation for tying. A small fire lit under the sheet rendered the sappy bark pliable, and at the appropriate moment it was turned inside out and the ends bunched and securely tied with bark cord. Thin saplings inserted at regular intervals across the canoe maintained the shape of the bark to the desired width. Cord ties held the spreaders in place. Further strengthening was achieved by forcing pliant branches into the body of the craft to act as ribs. The tied bark canoe, although an unsophisticated type, was a decided improvement on the simple Murray Valley form because its greater depth and rigidity increased seaworthiness although it too became waterlogged after prolonged use. (Smyth, 1878, Vol. 1, pp. 407-410).

More advanced types of boats were encountered on the tropical northern and Queensland coasts. These were of two basic types, the sewn bark canoe and the dug-out canoe, and were superior to those of the southern waterways. They allowed fishermen to venture out to sea for long periods. (Roth, 1910, pp. 6-17).

The sewn canoe was constructed from broad strips of bark obtained from one or two species of eucalypt from whose trunks they were easily peeled in the wet season. In some instances only a single sheet of bark was used, but this

This unusual floating wurly was a primitive watercraft seen on the Murray River about 1844

was unusual. Most canoes were built by extending one piece of bark length-wise from bow to stern along one side, and sewing two sheets together to form the other side. The two sides of the canoe thus formed were joined along the bow, stern and keel and then caulked with gum. To give added freeboard, the bow and stern were raised by sewing more strips of bark onto the sides. A network of ties, stretchers, braces and ribs was employed to maintain the shape of the boat. Small poles of mangrove wood were lashed along the gunwales to prevent the sides from collapsing. When completed, pieces of bark were laid along the bottom of the canoe to give added strength and to afford a dry, protected floor.

A superbly made sewn canoe, collected in Arnhem Land by the anthropologist Sir Baldwin Spencer during his 1901-02 expedition to North Australia, is displayed in the National Museum of Victoria. This example, measuring five metres in length, was brought to the mainland by eight Aborigines from the Pellew Islands and navigated up the Macarthur River for a distance of over eighty kilometres. Though the islands provided some shelter about thirty-two kilometres of open sea were negotiated during the voyage. (Spencer and Gillen, 1904, pp. 680-682, fig. 244).

The use of the dug-out canoe in northern Australia and Queensland was a result of Macassan influence during the regular visits by the trepang fleet out of the Celebes over several centuries. The Aborigines were quick to note the advantages of the sturdy log canoes carried by the Bugis and Macassarese fishermen and obtained them initially by trade or theft. Soon a new industry developed in the tropical forests and swamps of the north, aided by the metal axes obtained from the seasonal visitors.

The trepangers were responsible also for the introduction of the mast and pandanus sail, which became integral accessories both of dug-out and sewn bark canoes and which are so commonly represented in the cave art and bark paintings of Arnhem Land. (Davidson, 1935, p. 57).

To build a dug-out canoe, the Aborigines felled suitable softwood trees; usually the coral tree, *Erythrina indica*. This was roughly shaped and hollowed out on the spot with steel axes, and the partly finished canoe was then carried or dragged to the water's edge. Here, the final shaping was undertaken. Careful chipping was necessary to work the sides down to a uniform thickness of about two to three centimetres. The thickness was determined by tapping the canoe with a stick or stone, or by using the palms of the hands as a "precision gauge", placing one on each side of the area to be tested and moving them over it to judge its thickness. There was no keel on a dug-out canoe and its bottom was finished to a thickness of about eight centimetres. Dug-outs could be six metres long and about three quarters of a metre wide. (Tindale, 1926, pp. 104-109).

The use of an outrigger in association with the dug-out canoe is a com-

Tied bark canoes on Lake Tyers, Gippsland, Victoria, about 1888

This bark painting depicts a Malay Prau. Although it is over fifty years since the Malay Praus visited Groote Eylandt, the Aborigines have retained an accurate memory of their general shape, the double steering paddles, and the tripod mast

Making a dug-out canoe in the forests of Arnhem Land, Northern Territory

Aborigines and their bark canoes at Port Jackson, New South Wales, in 1790

paratively recent innovation. It was used widely on the Queensland coast where it had been introduced from New Guinea by way of the Torres Strait Islands and Cape York Peninsula.

The Aborigines along our tropic coasts obviously learnt from their contacts with Melanesian and Indonesian navigators, but the indigenous groups along the stormy southern shoreline did not have this advantage—nor did they have trees suitable for the construction of seaworthy log canoes. The large softwoods of the tropics were not to be found along the southern coasts or rivers.

In 1802 Matthew Flinders explored Australia's south coast. (1814, Vol. 1, pp. 53-193). At King George Sound, Western Australia, he saw neither canoes nor trees from which bark had been taken for making them. At Port Lincoln, South Australia, despite its fine harbour and numerous islands, he saw no sign of canoes. Possibly the Port Lincoln area never grew trees of the types which are suitable to provide bark for canoes. But the rivers and lakes of the Murray watershed and drainage system had an abundance of large trees and expanses of calm water, and this is where bark canoes were so common. Explorers encountered canoes frequently in this region and found that the Aborigines invariably took to the water in times of alarm. On one occasion, near Swan Hill in Victoria, a large group boarded twenty-four canoes and fled across a lake to a small island where they took refuge from an exploring party. (Mitchell, 1838, Vol. 2, p. 141).

Captain Charles Sturt, who discovered and explored the entire length of the Murray River early in 1830, observed on one of his first encounters with the aquatic groups (Sturt, 1834, Vol. 2, p. 201) that: "The natives launched their bark canoes, the only frail means they possess of crossing the rivers with their children. These canoes are of the simplest construction and rudest materials, being formed of an oblong piece of bark, the ends of which are stuffed with clay, so as to render them impervious to water."

Despite their simplicity of design, Europeans found it difficult to make bark canoes without instruction from Aborigines. In October 1818, the explorer John Oxley made a tied bark canoe to fulfil a desperate need to cross a stream in coastal New South Wales. He provides an interesting account of the occasion (Oxley, 1820, pp. 332-333): "In our route we had disturbed a large party of natives, some of whom were busily employed in preparing bark for a new canoe. There were several canoes on the lake, in which they all fled in great confusion; leaving their arms and utensils of every description behind them. One of the canoes was sufficiently large to hold nine men, and resembled a boat; of course we left their property untouched, though we afterwards regretted we did not seize one of their canoes, which we might easily have done. We however determined to send back in the morning for the unfinished canoe, and try our skill in completing it for use.

The Murray River above Moorundie in 1844

The Government post at Moorundie, on the Murray River, in 1842

Group of bark canoes on the lower Murray River, about 1860

"A party was sent back early this morning to secure the canoe, while we examined the river. The people returned in the course of the forenoon unsuccessful, as the natives had removed it with all their effects in the course of the night, throwing down and destroying their guniahs or bark huts."

"On Friday we returned to the entrance of the haven, and immediately commenced our endeavours to construct a canoe: our first essays were unsuccessful, but by Saturday night we had a bark one completed, which we hoped would answer our purpose; though I think if the natives saw it they would ridicule our rude attempts. This morning, the ebb tide answering, we commenced transporting our luggage, and in three hours everything was safe over."

The explorer Hamilton Hume (1855, p. 15) also attempted to make a canoe. On his way from Lake George to Port Phillip in 1824, he was determined to cross the flooded Murrumbidgee River, so he searched for a sheet of bark suitable to construct a canoe such as the Aborigines used. After much trouble, which he attributed to the lateness of the season and the fact that the sap was down and the bark firmly fixed to the trunk of the tree, he obtained a length of bark and succeeded in fashioning a canoe. But, to his great disappointment, it cracked and became useless.

A rare photograph of Murray River Aborigines painted with pipeclay in preparation for a ceremony

Bark canoes were seen wherever there was contact with the river people. George French Angas, roving artist and chronicler of the early 1840s, and son of one of South Australia's founders, George Fife Angas, included two bark canoes in his superb watercolours of Aboriginal life (1846, Pls. 30 and 47) and also recorded a graphic verbal description of the simple craft (Angas, 1847, p. 54): "I crossed the river [Murray] in one of their canoes; which are made merely from a sheet of bark from the gum tree, warped up at the sides by the application of moisture and fire, and stopped at the ends with strong clay. They are paddled by means of a long spear, having a sharpened kangaroo-bone fixed at one end, for spearing fish. The spear is held in both hands, and the paddler wields it standing; preserving the most delicate balance, which a breath of wind is sufficient to upset. During cold weather a fire is invariably carried in the canoe, raised on a small platform of clay, supported by wet weeds and mud; and by these fires they frequently cook a portion of their fish whilst on the water. Two, or, at the most three individuals, can be conveyed in the frail shells of bark."

Some of the most descriptive accounts of Aboriginal life in the Murray Valley were written by Edward John Eyre, the famous explorer and South

Australia's first Protector of Aborigines on the Murray River. He lived at Moorundie, below Blanchetown and 129 kilometres east of Adelaide, from October 1841 until November 1844. "A large canoe," Eyre (1845, pp. 264-265 and 314) wrote, "will hold seven or eight people easily; it is often twenty feet [6 m] long. The following is a description of an ordinary one for fishing:— length fifteen feet [4.57 m], width three feet [.91 m], depth eight inches [20.32 cm], formed out of a single sheet of bark, with one end a little narrower than the other and pointing upward. This end is paddled first; the bottom is nearly flat, and the canoe is so firm, that a person can take hold of one side, and climb into it from the water without upsetting it. It is paddled along with the long pine-spear *moo-aroo*, described as being used in fishing at night by fire-light. In propelling it the native stands near the centre, pushing his *moo-aroo* against the water, first on one side and then on the other; in shallow water one end of the *moo-aroo* is placed on the bottom, and the canoe pushed along."

"It is used too for spearing fish by night, which is by far the most interesting method of any. Having previously prepared his canoe, straightened his spear, and hardened and sharpened the points of the prongs, the native breaks up his firewood in small pieces, and loads his canoe with a stock calculated to last the time he intends to be absent. An oval piece of bark, about three feet [.91 m] long and two feet [.6 m] broad, is then coated over with wet mud and placed in the stern of the canoe, on a framework of sticks. One or two sticks are stuck upright in the mud, and others placed around them in the form of a cone. A fire is then put underneath, and the native, stepping into the bow of his canoe, pushes steadily into the stream, and commences his nocturnal employment. The wood of which the fire is made is of a particular kind, and, as only one description of tree [false sandalwood] will answer, it has frequently to be brought from a considerable distance. It is obtained among the brush of the table-land stretching behind the valley of the Murray, on either side, and its peculiarities are that it is light, brittle, and resinous, emitting when burning a most agreeable fragrance and a powerful and brilliant light, almost wholly free from smoke. Two men usually accompany each canoe, one to attend to the fire, and keep it always burning brightly, and the other to guide the canoe and spear the fish. As soon as the fire begins to blaze up the scene becomes most beautiful. The low black-looking piece of bark floats noiselessly down the middle of the stream, or stealthily glides under the frowning cliffs, now lit up by a brilliant light. In the bow is seen the dark, naked, but graceful form of the savage, standing firm and erect, and scarcely seeming to move, as with the slightest motion of his arms he guides the frail canoe. His spear is grasped in his hand, whilst his whole attitude and appearance denote the most intense vigilence and attention."

Pleasure craft such as this at Nildottie, South Australia, now use the river which knew the romantic era of the paddle steamers

Periodic flooding of the Murray-Darling Basin made canoes essential. This photograph was taken at Renmark in 1931

In times of drought the Aborigines abandoned their bark canoes. Low river at Waikerie in 1916

According to several early observers, harpooning fish with the long propelling stick was by no means solely a night activity, Beveridge (1889, pp. 63-64), a pioneer settler in the Swan Hill District of Victoria, observed: ". . . they select a stretch of shallow water, full of reeds and other aquatic plants, over which the wary fisherman propels his canoe, using the plain end of the stick for the purpose. Every now and then he thrusts the stick sharply to the bottom, thereby disturbing the feeding fish. As a matter of course they rush away from the disturbance, shaking the plants in their hurry, which at once tells the keen-eyed fisherman the position of his prey. After the plants have ceased shaking, the wily savage pushes his canoe gently up to within striking distance of the plants which were last in motion, he knowing right well that at the foot thereof his game is resting. Poising his grained weapon but for a short space, he launches it with precision, and seldom fails to bring his scaley victim, quivering and glittering to the surface.

"When sailing over deep water, both ends of the stick are used; it is held by the middle then, and each end is dipped into.the water alternately."

J. C. Hawker, a South Australian pioneer who arrived in October 1838

Bullock team makes an easy crossing of the Murray River in the midst of drought. The construction of locks at regular intervals now maintains the river level

on the *Pestonjee Bomanjee* with the second Governor of South Australia, George Gawler, made use frequently of bark canoes while living near Moorundie in the early 1840s. He provides some vivid accounts of their use as an extension of the hunter-gatherer life of the Aborigines. One of these reads: "Went up in the canoes to Piomubko, a lagoon about a mile [1.6 km] from the Pound and which by the flats being under water is now joined to Cumpungo, our natives speared two fine harbet in this lagoon. We got the canoes into the river by means of a creek which was from the river to this lagoon, and then proceeded down the river to our camp at Cumpungo, on our way the natives caught some turtles that were lying basking in the sun on the stumps of trees which were near the water. Their manner of catching turtle is ingenious, the moment they see the turtle, they shove the canoe into, or near the reeds off the banks of the river and without noise they slip quietly over the side of the canoe into the water they dive and come up with a piece fitting exactly like a green wig on their heads, they then float noiselessly down till within a few yards of the log on which the turtles are and then diving again the weed floats down with the stream and whilst the turtles are watching it the native suddenly rises close to the log and catches one or two as they leap into the water. We got a good number of turtle eggs they are about the size of a pigeons and eat like a plovers." (Hawker, 1841-45, p. 85).

The Aborigines of Lakes Alexandrina and Albert, on the lower Murray, were particularly adept at spear fishing from a canoe as the shallow waters of the lakes were heavily stocked with fish which were an important part of the local diet.

This form of hunting required both patience and skill and on occasions during summer months one group would challenge another to a competitive hunt on some favoured ground. Several days before a contest a large assembly would gather at the appointed place. The two groups selected a number of their best huntsmen who were introduced at a "canoe corroboree" on the evening before the competition. Canoes and spears were prepared before dawn and then competitors and spectators would make for the hunting ground in their bark canoes. They ringed the appropriate site while the challengers took turns at spearing fish aroused for the purpose by a selected canoeman. After two days of keen competition the most skilled huntsman was chosen and without any outward sign of elation he would be towed to the camping ground where he was received with excited approval by all concerned. (Smith, 1930, pp. 231-236).

These competitions were times of great enjoyment and occasions for meetings between different river groups. Berndt (1941, p. 24) relates how the Moorundie people travelled down the river for such ceremonies, carrying possum and kangaroo skins and kangaroo sinews and bones for use in an exchange of gifts.

The canoe ceremonies of the lower Murray, said to be the most graceful of the traditional corroborees, extended upstream along well defined mythological lines. Smyth (1878, Vol. 1, pp. 174-175) describes a "canoe-dance" at Moorundie, and it was also one of the principal rituals performed by the Rufus people of Lake Victoria. Both men and women joined in this ceremony, and were painted for the occasion with white and red ochre. The performers are said to have formed themselves into a double row, each person with a stick held behind his arms. They moved their legs alternately to the time of the song. Suddenly and simultaneously they would all remove the sticks from behind their arms, and hold them up in front, and then commence swaying their bodies alternately from side to side, in the most elegant manner, imitating in all their movements the paddling of their bark canoes. (Angas, 1847, pp. 101-102).

The main legendary hero of the lower Murray people (Jaralde) was named Nurundere or Ngurunderi. He was attributed with the formation of the great Murray Valley in the Aboriginal Dreamtime. According to Taplin (1874, p. 42): "He is said to have made all things on the earth, and to have given to men the weapons of war and hunting. Nurundere instituted all the rights and ceremonies which are practised by the Aborigines, whether connected with

A picturesque canoe tree near Giles Creek, a tributary of the Finniss River, at Kondoparinga, South Australia life or death. On enquiring why they adhere to any custom, the reply is, because Nurundere commanded it."

"He is represented to have been a great hunter, and there were contemporary with him two other remarkable hunters named Nepele and Wyungare. According to the natives they must have been a mighty race, and the game which they pursued gigantic, for the salt lagoons are the places where Wyungare and Nepele used to peg out the skins of the immense kangaroos which they killed, and thus denuded them of grass. A mound on the Peninsula is still pointed out as the remains of the hut of Wyungare."

Tindale (1935, p. 273) states that two great fixed sandhills at Mount Misery, beside the main road at Section 219, Hundred of Malcolm, represent the canoes of this ancestor, lying still in the position in which he left them.

The most comprehensive available account of the Ngurunderi myth was recorded in 1940 by Professor Ronald Berndt (1964, pp. 203-204) from the late Albert Karloan, who was then aged seventy-five years and was the last surviving fully-initiated man of the Manangri clan belonging to the Jaralde. "The ancestral hero Ngurunderi paddled his bark canoe down the small creek which was later to become the River Murray. He had come from the Darling, following the giant Murray cod. As this fish swam, its tail swept aside the water, widening the river to the size it is today. When Ngurunderi paused to rest the cod swam on into the Lake, and he gave up hope of catching it. Then he thought of his 'wife's brother,' Nepele. Quickly getting into his canoe he rowed to Bumondung, and from there called out to Nepele, who was sitting on a red cliff named Rawugung, Point McLeay. Nepele pushed out his canoe, rowed it to some shoals, and waited with spear in hand. The cod swam down towards Nepele, who speared it opposite Rawugung and placed it on a submerged sandbank there. When Ngurunderi arrived they cut up the cod into many small pieces, throwing each into the water and naming the fish it was to become. Finally they threw the remaining part into the lake saying 'keep on being a Murray cod.'

"Ngurunderi continued his travels. Eventually he reached Bamundang, where he disembarked and pulled up his canoe: his footprints are still there. Carrying the canoe he walked to Larlangangel, where he left two large mounds of fresh-water mussels. One day, on his way back from Granagung, he saw some people at a place called Ngirlungmurnang. They were frightened of him and hid in the reeds. But Ngurunderi could hear them whispering, and he transformed them into a species of blue birds. At this juncture Ngurunderi's two wives appeared. They were at Gurelbang cooking the *dugeri* (silver bream), tabu to women, and the breeze blowing from that direction carried the smell to him. Having no further use for his canoe, he stood on the two mounds at Larlangangel, and, lifting it up, placed it in the sky where it became the Milky

"Pelican," described by the Reverend George Taplin as a "thorough savage," holds a heavy spear of the type used both for fishing and to propel his bark canoe over the waters of Lakes Alexandrina and Albert in about 1860. He died at Point McLeay, South Australia, on 19 August 1871

Way. He then set off for Gurelbang. In the meantime the two women, thinking Ngurunderi might smell the fish, had made their escape on a reed-raft, poling their way across Lake Albert to Thralrum, on the western side. There they left the raft, which was metamorphosed into the reeds and yaccas found at that point today, and continued down into the Coorong.

"When Ngurunderi reached Gurelbang and found them gone he too made a raft, and followed them into the Coorong. Here he met a malignant spirit named Barambari. Ngurunderi asked whether he had seen the two wives, but Barambari started a quarrel and speared him in the thigh. Ngurunderi laughed, pulled it out and threw it away. Then he threw his club, knocking Barambari unconscious, and thinking he was dead turned to go. But Barambari regained consciousness, and manipulated his magical spearthrower in such a way as to stop Ngurunderi from walking on. Ngurunderi returned, and killed him with his club. He lifted some large gums and other trees, piled them into a heap and set them alight, then lifted Barambari's body and placed it on top of the blazing pyre so it would be completely consumed. Turning around he tried again to walk away, but again could not do so. He picked up all the congealed blood and threw it on the fire, and after that he was able to continue. At Wunjurem he dug a hole in the sand to get fresh water. Kneeling down to drink he put his head against the sand, and this depression was transformed into rock.

"Eventually he came to Ngurunduwurgngirl ('Ngurunderi's home'), where he lived for some time, giving up all hope of finding his wives. Later he continued his wanderings down the coast along Encounter Bay, and after a number of adventures was about to cross over from the mainland to Kangaroo Island when he saw his wives starting to do so. It was possible, at that time, to walk across to the island. When they reached the centre Ngurunderi called out in a voice of thunder saying, 'Fall on them, you waters!' Immediately the sea rose, and they were drowned; but they were metamorphosed into Meralang, 'Two Sisters,' now called The Pages, north-east of Cape Willoughby on Kangaroo Island. Ngurunderi then went to Kangaroo Island, called Ngurungaui, meaning 'on Ngurunderi's track,' referring to the path taken by all spirits on their way to the spirit world. He made a large casuarina tree, under which he rested. Then he walked down to the western side of the island, and threw away his spear into the sea: rocks came up at that place. Finally he dived into the sea to cleanse himself of his old life, and went up into the sky: Waieruwar, the spirit world. But before disappearing, he told the Jaraldi people that the spirits of their dead would always follow the tracks he had made, and eventually join him in the Sky-world."

The Aborigines were experts at converting the few basic raw materials in their environment into useful weapons and utensils. Probably the idea of

Teenminne, wife of "Pelican," with her young child and dressed in a warm skin cloak. After her death on 21 September 1869, the Reverend George Taplin wrote, "She was my first friend among the natives and was a truly excellent woman; kind hearted, intelligent, and faithful."

A unique photograph of a Murray River bark canoe being made in 1862

dug-out canoes did not occur to them, but they had devised a simple craft from sheets of bark capable of meeting the requirements of their environment.

A young German traveller, named Gerstaecker, who undertook a journey from Albury to Adelaide while on a visit to Australia during a world tour in 1851, demonstrated that the trees of the Murray Valley were unsuited to the manufacture of dug-out canoes. As no craft was available he resolved to make a canoe but found the local timbers unsuitable: ". . . nothing but gum-trees, with only a few stringy-barks—no pine, no cedar. There were pines, I was told, farther down the river, in some places, but none large enough to dig out a good canoe." (Gerstaecker, 1853, Vol. 2, p. 317). A search of the adjacent stringy-bark forests failed to locate a suitable log and he finally resorted to a heavy, unwieldy river gum. It took several days of constant chipping to hollow out the trunk to form a canoe which, when finished, proved very heavy and sank low in the water. Gerstaecker (1853, Vol. 2, p. 325) considered it a poor craft compared with the canoes he had used in North America. These were ". . . light craft, cut out of a pine and perfectly seasoned, shooting like an arrow rather over than through the waters."

The young adventurer had not travelled far when he encountered a party

A corpse covered with leaves is followed by a funeral procession of Aborigines in bark canoes on the lower Murray River in 1864

of Aborigines who were astonished at the sight of the clumsy log canoe. Gerstaecker (1853, Vol. 2, p. 328) comments: "... their own bark-canoes [are] only pieces of peeled bark lying flat on the water, and very seldom larger than would carry a black with his spear, or at the most two men."

After a precarious journey of 130 kilometres the canoe was abandoned and the long and adventurous journey to Adelaide completed on foot.

The time required to make a bark canoe depended upon several variables: its projected size, the number of people available to assist, the season of the year, and how urgently it was needed. If the occasion warranted, a piece of bark could be removed from a river red gum with the aid of a stone axe and converted into a small canoe in a matter of half an hour. Frail and perishable no doubt, but adequate for immediate use. Hawker (1841-45, p. 96), who used bark canoes almost continuously while at Moorundie, had no difficulty in obtaining canoes at short notice in April 1844. He instructed an Aboriginal to cut two small canoes, and returned on the following day to find them completed. The procedures involved in making these craft must have been expertly carried out, because he and a companion used them on the next day to go out on a lagoon and shoot teal and geese.

Such small canoes could be made rapidly, but a number of men was required to make a large and durable canoe. After it had been shaped and tied, a seasoning period of two weeks and regular rubbing with grease and ochre were desirable if the canoe was to last for its maximum life of about two years.

Easy removal of large slabs of bark was possible only during the late spring-summer growth period, when the sap flowed freely between the bark and the hardwood core of the tree. So this was another determining factor in the time taken to make a canoe. The period of the year most suited to bark stripping for canoes varied according to the season. The sap was not rising freely in trees along the Murrumbidgee River in October 1824 when Hovell and Hume wanted to obtain bark for a canoe (Hovell and Hume, 1831, p. 8) nor on the Darling River in October 1844 when Captain Sturt required a canoe for one of his party to cross the flooded river. Cambọli, his Aboriginal companion, is reported to have had a "fatiguing task" in removing bark for a small canoe. (Sturt, 1849, Vol. 1, p. 138).

Hawker (1841-45, p. 79) appears to have had no difficulty in having canoes cut in June, August and September 1843, and in April 1844, but Le Souef in his description of canoe making at Swan Hill in 1848 (undated MS., p. 25), describes how the Aborigines had to strike the bark all over in order to loosen it. It is interesting to note that Aborigines in other parts of Australia used the same techniques and encountered the same problems when removing pieces of bark for shields and carrying dishes.

Le Souef tells how: "The blacks on the Murray used to cut very large canoes from the old Red-gum trees growing in the bends and banks of the river. I have seen eight or ten men engaged in cutting one. First the shape was marked out on the tree chosen: while two or three cut the bark, others tapped it all over with their tomahawks to loosen the bark; others skilfully drove small bark wedges in and worked the bark off with long saplings brought to a flat point. Before it was quite separated from the tree, ropes were passed round it to prevent it from falling and it was then allowed to slide quietly to the ground. The fastenings were then slackened, and the men, going to each side, finally laid it on the ground. Then two or three pointed sticks were placed across the bark to keep it from curling and it was then washed with thin liquid clay to keep the sun from cracking it. A small bank of clay was made in the bow, on which a fire was kept burning, either at night, when fishing, or on a cold day. Some of these canoes carried a dozen or more people, and with care they lasted for months."

The trees sought to provide bark for canoes varied with the district and also with the size of the canoe required. Heads of families generally needed canoes sufficiently large to convey their families and possessions, while single men contented themselves with smaller canoes, finding them more suited to pursuing aquatic birds during the moulting season. Generally, a two metre

This picture of fishing canoes on the Murray River shows how the Aborigines used their heavy spears to propel these simple craft

canoe would carry two people: one of two-and-a-half metres in length would take three passengers, and a three to three-and-a-half metre craft sufficed for four travellers. Large canoes of up to six metres could comfortably accommodate six or more people.

The best canoes undoubtedly came from the river red gum with its thick bark, but the mountain ash, box, stringybark, some blue gum, and the white gums of the river valleys, and even Snowy River mahogany, were used, particularly in Victoria and New South Wales. The red gum was plentiful along the Murray, and the most suitable trees had slight curves in their trunks. Bark selected from such trees saved the Aborigines many hours of work, because if they used bark taken from a straight trunk, they had to curve each end by means of fire.

Hawker (1901, p. 3) recorded a comprehensive description of the river canoe-maker's techniques which he must have witnessed on many occasions: ". . . a good deal of judgement was required in the manufacture. Several trees were carefully examined for the thickness of the bark — the gum tree is nourished entirely by the sap running through the bark. The centre of the tree may be burnt by bushfires, but so long as a strip of bark remains there will be foliage.

Removal of a large sheet of thick bark for a canoe left the symmetrical scar on this red gum at Milang, South Australia

Opposite:
Many centuries ago, the Aborigines cut bark for a canoe from this fine red gum at Nor-West Bend Station, near Morgan, South Australia

The gum trees along the banks of the Murray were well nurtured owing to their roots being watered sometimes by the overflow of the river, and always by the dampness of the soil a short depth below the surface: consequently the bark was much thicker on some trees than others, and therefore unserviceable for making a canoe. In testing a tree the native took a yam stick, made of hardwood with a flat point. With this he began making cuts in the bark for his toes, and also to hold on by when using his yam stick; and by this process he went up trees not only for canoe-making, but also for birds' eggs or opposum. Having found a suitable tree, he, when about 20 ft. [6.09 m] from the ground, commenced cutting the shape of the canoe through the bark to the hardwood, with his yam stick. This process completed, a rope was passed around the tree, taking about the centre of the canoe. This was to prevent the piece of bark from slipping off before the proper time, in which case a smash would be inevitable. The next proceeding was a ticklish one, and the fate of the canoe depended on the care bestowed on it. Several pliable, thin, flat sticks were carefully forced in between the bark and the tree; short spaces apart, but only short distances at a time. When a large canoe was being made a second cord was passed around the tree, so that the canoe should not tip over when separated from the tree. As soon as the separation was performed the cords were eased so as to allow the canoe to slip down until the lower end rested on the ground, and six to eight natives were ready to receive it as the cords were gradually loosened, and place it flat on the ground. Now commenced the manipulation of this great, heavy piece of bark into the proper shape for a canoe. The bark at the upper end was much thinner than that at the lower end, the latter being nearer the base of the tree. Props were now placed under the sides of the canoe, and small fires were made inside at various distances to evaporate a good deal of the sap, and also to cause the sides to curl up sufficiently to keep the water out when afloat, stretchers being put across to prevent too great a curl. Much care had to be used in the drying and curling process, as too great or sudden heat would split the bark and render it useless, but this seldom happened. [Another method sometimes used to shape bark canoes has been recorded for the Wellington region in South Australia. Here, on occasions, the Aborigines scraped out a hollow in the ground to the desired shape of the canoe. The freshly cut bark was placed over the hollow and weighted with mud and clay and then left to dry to the ideal shape]. The thin bark at the upper end was made the bow and bent well up, while the thick bark at the lower end represented the stern, and this could not be curled. In order, therefore, to keep the water out a mass of clay from the river bank, well mixed with grass to bind it, was put as a barricade against the water; and if renewed occasionally, it served the purpose very well. Several days were required before the canoe was sufficiently dried to put in the water, and even then the weight was so

Opposite:

Centuries of growth have almost healed the large scar left by removal of bark from this ancient canoe tree at Chowilla, South Australia

Aboriginal bark canoe on the Wimmera River, Dimboola, Victoria

Aboriginal and child in a bark canoe on the Murray River in 1907

great that several natives had to drag it down to the river; but when once launched, it lasted for a considerable period. To propel the canoe a thinned sapling fir tree about fourteen feet [4.26 m] in length was used, to one end of which were attached two sharp mallee-wood prongs for the purpose of spearing fish, one prong being a little longer than the other. This end of the pole was also employed for paddling, while the other served to push the canoe along when in shallow water. In cold weather a clay coating was put on a small space in the bottom of the canoe, and a fire was lit for warmth. The ropes used for holding the craft to the tree were made by the natives from a very fibrous grass, and so were all the cords and the twine for making their nets. When making the fires inside to evaporate the sap, great care had to be used to keep the bottom from curling, and to ensure its being quite flat, stones were placed to weight it down. The canoe to a Murray River native was as useful as a vehicle to a white man by its means he was able to move his family and small belongings to many localities along the river, as well as frequently provide sustenance for himself and them by fishing with lines and spearing fish in shallow lagoons."

"Old Malo, my blackfellow, succeeded in cutting me a good canoe out of a tree about 5 miles [8 km] from my place, adjacent to a lagoon called Cumpungo. I went up there as soon as it was ready for launching, and he superintended the performance, which proved quite satisfactory. I came down in the craft to Moorundie, Malo giving me a lesson in paddling, which was always done standing a little behind the centre of the canoe, except when after game—then a squatting position was assumed."

Although bark canoes were capable of transporting heavy loads, they required a fine sense of balance and agility in handling—especially the smaller types. These skills were imparted during childhood when the young Aboriginal began his training in the arts of hunting, food gathering, dancing, singing, weapon making, and all the aspects of their ancient culture.

A number of sources testify to the skill and nonchalance with which the Aborigines navigated their canoes. "They have a wonderful knack in their management of canoes, driving them along with amazing velocity, and a directness of bearing that would delight the very heart of a pilot whose pride it is to con a ship successfully," commented one Murray River pioneer. The canoeman did not bring his craft end-on to a sea, but bow-on, so as to sidle over the waves, the canoe riding over sideways like a duck. It would probably break its back across the wave if faced end-on (Smyth, 1878, p. 411). The slightest loss of equilibrium by any of the occupants was likely to upset the frail bark and throw them and their cargo into the water.

Le Souef (undated MS., p. 25) had occasion to be thankful for the Aboriginal's skill in the management of canoes when crossing a wide stretch of flooded

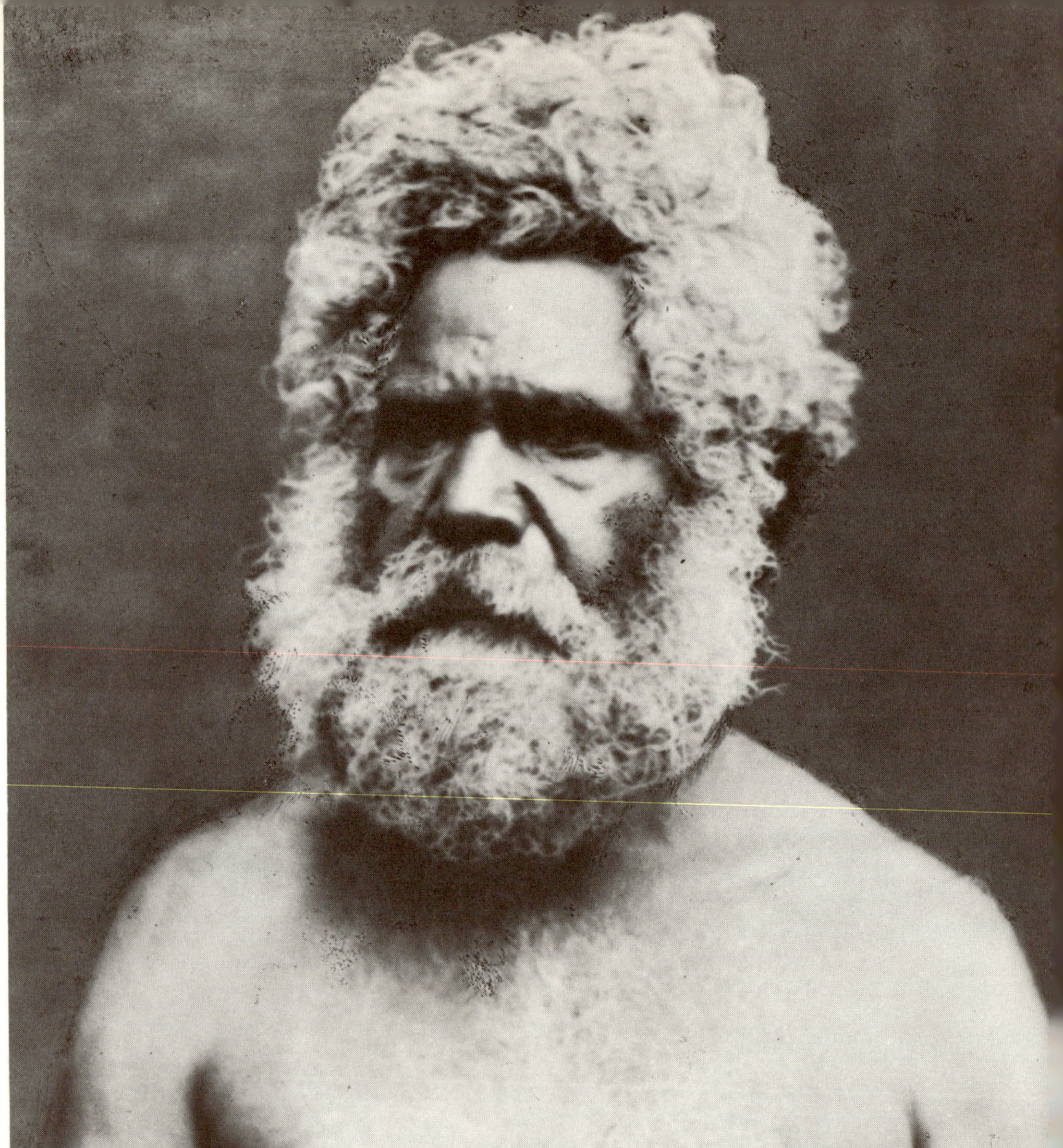

"River Murray Dan," one of the last survivors of an Upper Murray tribe, photographed about 1890

river above Swan Hill to reach Murray Downs Station: "Robert Peel, who had a large canoe, offered to take me over for a fig of tobacco, and although I did not like him, I knew he was a good hand in a canoe and so accepted his offer. The distance we had to go was over two miles [3.2 km], and when we were about half way, it came to blow, so that quite a little sea arose. I thought we should have been swamped, for a canoe, being only formed of a single sheet of bark, has very little buoyancy, but I certainly never saw a man before, or since, handle a canoe as did this black. I baled out the water as well as I could, for a good deal came over the side, in spite of all the darkey could do, but he got safely across, and I gave him two figs of tobacco instead of one. I stopped with my friends the Jamiesons until next day, as I had had enough of the river for that night."

An example of how the canoes could come to grief if their crewmen lacked skill or caution, was seen during the pioneer river journey of the *Lady Augusta* when the following incident occurred on 6 September, 1853: "As we approached the Darling we were hailed by a party of natives. It was quite dark, but they had a fire which threw out its light to a considerable distance. Two men and a woman came alongside in a canoe, but managed the matter so awkwardly that it was swamped under the *Eureka*'s stern, and they all fell into the water. The men ran up [the side of the boat] like cats, and the woman was following, but they kicked her down again, and she fell back into the river with a piteous scream. It is probable Anglo-Saxon gallantry might have attempted a rescue, but an instant showed that it was not required. She struck out bravely, and the man assured us she would soon reach the shore. They had thrown her into the river with the hope of her recovering the canoe; but in this they were disappointed. It was a case of total loss, and the underwriters, if any, will have to make it good." (Andrews, 1854, p. 210).

This concern for the craft, though hardly chivalrous, was understandable. For all its crude form, the making of a canoe was no small task in the survival-oriented life-style of the Aborigines.

The Aborigines of the riverine areas held canoes in high esteem, because they extended their range of hunting and food gathering from the land to the extensive waters of the river systems. They did not carry the canoes with them on their constant wanderings, but when they left some particular camp or hunting place they concealed the craft very cunningly among reeds or bushes, where it lay secure until their return in the pattern of their travels.

Infrequent sightings in some areas led to a belief that the Aborigines did not make large numbers of canoes until steel tools became available. This is not the case, because their skill in canoe-making was highly developed long before they handled European tools. The tools themselves soon changed, but

The Aborigines climbed to a great height to obtain a suitable piece of canoe bark from this red gum on the Edson property at Blanchetown, South Australia

Large canoe trees photographed at Waikerie, South Australia, in 1914

not the techniques. On the other hand, it is possible that the production of canoes increased for a brief period after their introduction.

On the Darling River in 1844, Charles Sturt had assisted the canoe builders to move into the Iron Age. Upon meeting an Aboriginal working diligently with a stone axe on a riverbank tree, Sturt (1849, Vol. 1, pp. 127-128) recounted in generous fashion: "With such an implement he would hardly finish his

Large sheets of thick red gum bark suitable for making canoes were sometimes hard to obtain. This photograph of a canoe tree at Chowilla, South Australia, shows how the Aborigines had to climb to great heights to obtain suitable bark

Aborigines fishing from a bark canoe at Chowilla, South Australia, in 1886. "King" Billy (left) is seated in the canoe

work before dark. I therefore sent for an iron tomahawk, which I gave to him, and with which he soon had the bark cut and detached. He then prepared it for launching by puddling up its ends, and putting it into the water, placed his lubra and an infant child in it, and giving her a rude spear as a paddle, pushed her away from the bank. She was immediately followed by a little urchin who was sitting on the bank, the canoe being too fragile to receive him; but he evidently doubted his ability to gain the opposite bank of the river, and it was most interesting to mark the anxiety of both parents as the little fellow struck across the foaming current. The mother kept close beside him in the canoe, and the father stood on the bank encouraging his little son. At length they all landed safely, when the native came to return the tomahawk, which he understood to have been only lent to him. However, I was too much pleased with the scene that I had witnessed to deprive him of it, nor did I ever see a man more delighted than he was when he found that the tomahawk, the value and superiority of which he had so lately proved was indeed his own. He thanked me for it, he eyed it with infinite satisfaction, and then turning round plunged into the stream and joined his family on the opposite bank."

White settlers benefited from the availability of Aboriginal canoes. We know that Hamilton Hume recognized their potential when confronted with a flood. Major Mitchell used them on one occasion at least. Gerstaecker, on his trek down the Murray Valley in 1851, found a bark canoe invaluable to avoid a group of hostile Aborigines. He remarks (Gerstaecker, 1853, Vol. 3, p. 40): "I had to pass several mobs of blacks before I could reach the Darling . . . Once I had to take one of their bark canoes to pass a large mob, holding a wild corrobery on the banks of the Murray, while the dogs were out in the woods chasing opossums; and I did not dare to cross their track. I was obliged to glide afterwards down the stream in the dark, under the very bank upon which they were dancing."

Early squatters also used canoes to ferry precious sheep across swamps or streams on their pioneering treks. E. M. Curr, who settled on the Upper Murray in 1841, witnessed the removal of a large sheet of bark from a river red gum and noted the final preparations for a canoe and its use to ferry sheep across the river: "Once landed, it [the bark] was placed in the position of a boat turned upside down; some dried leaves and boughs were put under it and set fire to, the object being to dry up the sap, and so toughen the bark as to allow of its being moulded into the required shape. In a minute or two the fire had burned out, the canoe was righted and placed on a level spot; two or three pieces of stick were inserted after the manner of *thwarts* to keep the sides distended; the bow was permanently elevated, being forced into the required position by means of a log placed under it, and in this way the canoe is usually left for a day or two to set and harden. The canoe I speak of (which could have been used as soon as it came off the tree) was about 18 feet [5.45 m] long by 2 feet 6 inches [.76 m] wide, floated eight inches [.2 m] clear of the water, and would carry five persons."

"Having thus got a canoe, I set to work in the morning to erect yards at a convenient place on the bank of the river, and the next day, with the assistance of my men and the three Blacks, I got one of my flocks transported across the Goulburn. The operation was performed by placing the sheep, six at a time, in the canoe with their legs tied, when a Black punted them over. This wet and troublesome job accomplished, my men set vigorously to work to provide the first necessaries of a station." (Curr, 1883, pp. 90-91).

A party, overlanding a mob of cattle from Sydney to the lucrative market of the Victorian goldfields, was confronted with the flooded Lachlan River and enlisted the aid of the Aboriginal and his simple canoe: "I rode down the river some eight miles [12.8 km] to a camp of blacks we had heard of and I got three of them to return with me to cut a bark canoe. After a good deal of search about, we at last found a suitable tree and in a couple of hours had cut a good-sized canoe and dragged it nearly a mile [1.6 km] to the river's edge,

A canoe tree near the Hindmarsh River on the Hutchinson property, Hindmarsh Valley, South Australia

Ngulgare, photographed about 1860, wears one of the rugs which the Aborigines made from the dense, warm skins of the possums. He holds a parrying shield, and a reed spear in the throwing position

One of the Narrinyeri people who settled at the mission established by the Aborigines Friends' Association at Point McLeay, South Australia, in 1859

Moilge, photographed about 1860, dressed in animal skins and holding the typical reed spear and woomera (spear thrower) used by Aborigines of the Murray districts but though the blacks had helped us so far we could not induce them to cross the river. They said the current was too strong and made all sorts of excuses, and, on receiving the tobacco we had promised them took themselves off.

"My early schooling in the art of managing a bark canoe now again served me in good stead. I stripped off my clothes, excepting my shirt and trousers and took the canoe across. On my return, I loaded some of our belongings and got them safely across. As it was now late in the afternoon, we thought it best to defer the crossing until the following morning.

"Soon after daybreak, we commenced. I took over the rest of the stores and things we had on the dray, and then the bows, yokes and chains of our team; when this was safely accomplished, I had to take our bullock-driver (Brandy), over, and this was no easy matter, as he could not swim, also he was very nervous, and the least swaying motion in a bark canoe is fatal. I told him to kneel down and fix his eyes on the bottom of the canoe and on no account to look at the water. I also impressed upon him that if the canoe sank, owing to any stupidity on his part, I should be unable to help him, as I should have quite enough to do to look after myself. I then shoved off, and standing perfectly motionless in the centre, paddled across as quickly as I could. Brandy never spoke a word and kept perfectly still. When I reached the opposite side I held the canoe steady, while he got out.

"The next operation was to float the dray across—first sending to the other side a team of bullocks, which, when landed, were to be used to drag across, by means of ropes, the dray itself. We got eight steady bullocks and with the help of our whips, drove them into the river. They were, by the strength of the current, carried down a good way but got across alright . . . We next tied two empty casks, which we carried for the purpose, to the guard irons of the dray, over each wheel. Fortunately, we had provided ourselves with plenty of string clothes' line and heavy dray ropes in case of getting into a fix such as we now found ourselves in. Ogilvie [his partner in the venture] climbed into a gum-tree overhanging the river taking with him a ball of string, one end of which I put in my mouth, making a large knot at the end, the better to hold it, and then getting into the canoe, I started for the other side . . . I got three parts of the way over very well, but then the string unfortunately struck the water and I thought my teeth would have been pulled out, but I held on till within about twenty feet [6 m] of the bank, when the strain compelled me to let go or I should have been pulled backward. Ogilvie rolled up the string and managed to climb some fifteen feet [4.5 m] higher up the tree, and I then started again and got nearly across before the string touched water, but when it did, I felt as if my teeth and head both would be dragged off. I was, however, determined not to be beaten if I could help it, and struggled on close to the bank; Brandy then rushed in over his waist, clutched the canoe.

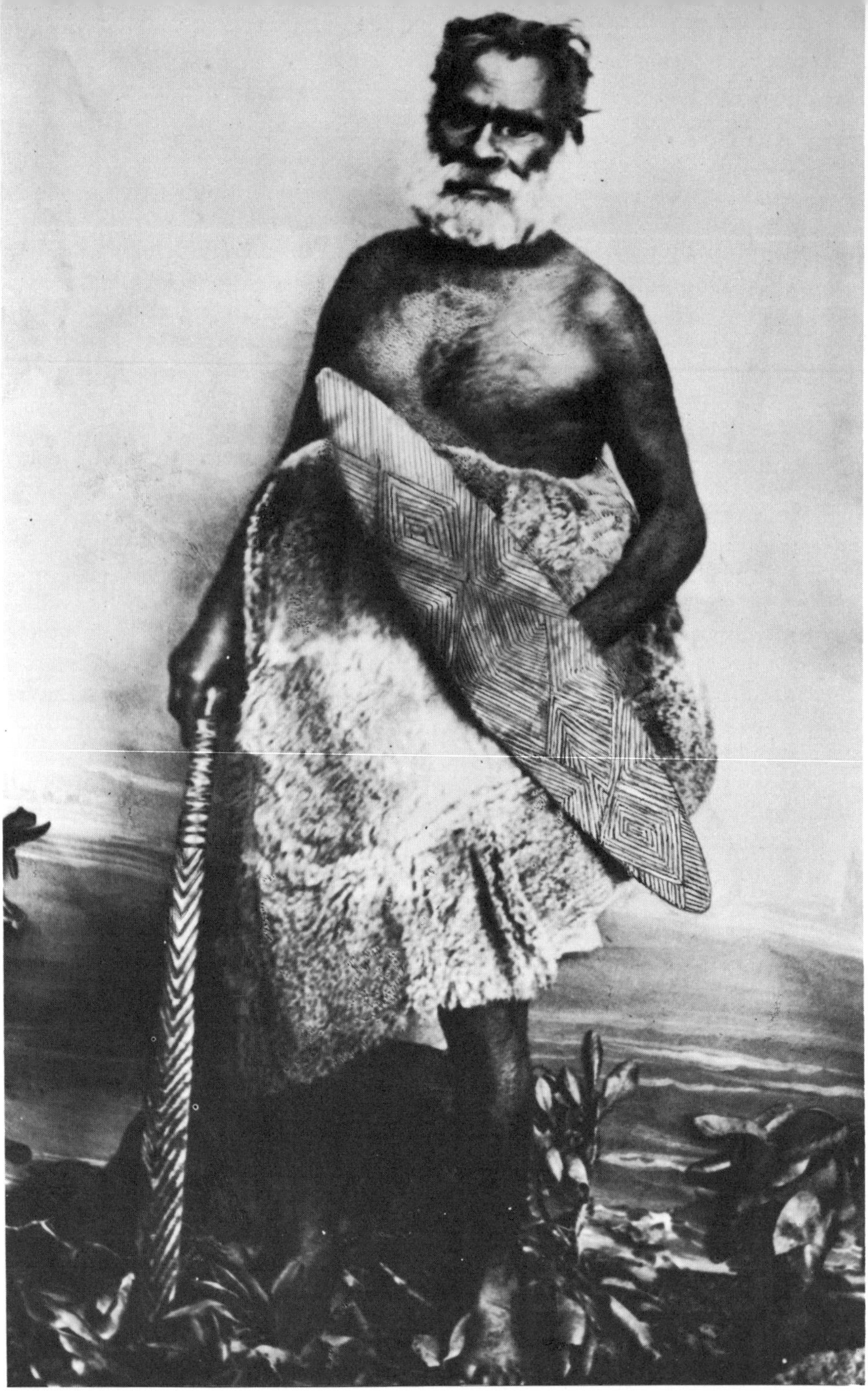

An Aboriginal of the Lakes Alexandrina and Albert region of the Lower Murray. Dressed in a skin cloak, he holds a hardwood parrying shield and a club. Taken about 1860

Opposite:
Three canoe barks were removed from these two red gums on the banks of the Murray River at Nor-West Bend Station, near Morgan, South Australia

This bark canoe, made at Avoca Station on the Darling River in 1904, is preserved in the South Australia Museum

As soon as he did so, and I had my arms free, I took the string in my right hand and jumped out and tied the end round a small sapling until Ogilvie got down from his perch and bent the string on to some clothes lines, these again being fastened to the dray ropes. Brandy and I then pulled the rope over and made it fast to the team, the other end being securely tied to the end of the dray pole; the bullocks then dragged it across without mishap.

"It was now getting late in the day, but we determined, if we could to cross

Opposite:
When the Aborigines of Chowilla, South Australia, cut a canoe bark from this tree they avoided the rough bark at the foot of the trunk

These two photographs of the same canoe tree at Blanchetown, South Australia, were taken with an interval of nearly eighty years. The small amount of new wood round the scar shows how slowly these trees heal themselves the cattle so, placing the team already crossed close to the water, where they could be seen by the others, we drove the mob down to the water's edge. To our great joy, an old bullock that had evidently been used to swimming, took the lead and struck out manfully, all the others following and we had the satisfaction of seeing them all land safely. Now there were only Ogilvie and Billington to take over. I was very tired and it was getting dusk, so I persuaded them both to get in. Before doing so, we drove into the river the horse on which Billington had been tailing the cattle all day and Brandy caught him as he landed. We now made what I hoped was the final trip, but I soon found that our frail craft was over-crowded, as the water began to come in over a low place on one side. There was a submerged branch of a tree in the river close by and with a few strokes of my canoe pole, I laid the canoe alongside and told Billington to jump on to it as quietly and quickly as he could; this he accomplished just in time and there he sat bobbing up and down on the branch until having taken Ogilvie over, I went back for him and got him safely landed just as it was getting dark." (Le Souef, undated MS., pp. 59-61)

The use of bark canoes by pioneer pastoralists was not limited to the transport of people and livestock. In the early years of the establishment of remote Mildura Station these seemingly fragile craft were used to ferry bales of wool, weighing 400 pounds [181.43 kg] apiece, across the Murray River from the Victorian side to the road on the opposite bank. (Allen, 1853, p. 29).

Even geologists resorted to canoes to aid their surveys. A. W. Howitt, when a warden of the goldfields of Victoria, descended the Mitchell River in east Gippsland in a canoe for the purpose of extending a geological survey of the region. "The canoes were no great triumph of naval architecture," he said in his report. (Smyth, 1876, pp. 214-220). Of his travel in a canoe of about two and a half metres long, which was the largest of two made especially for the journey, he recorded: "I sat down on a piece of bark which was luxuriously provided for my special use, with my geological hammer in one hand and my notebook in the other; I felt that I only wanted another pair or two of arms to present a ridiculous parody of a crosslegged image of Brahmah."

The bark canoe was not always looked upon as an aid to the settlers. Sometimes the canoes enabled the Aborigines, with their highly developed skill on the water and familiarity with the environment, to outmanoeuvre the invaders. When a punitive expedition was sent to the Rufus River region near Lake Victoria, in June 1841, to investigate an Aboriginal attack on a party which had been droving sheep overland to the infant settlement of Adelaide, the Aborigines took to the water to make their escape. Some swam while others crossed in eight bark canoes.

The only way in which the pursuers could follow them was by improvising a ferry. They lashed a tarpaulin round the body of a dray, and eighteen men

The steamer *Lady Augusta*, with her barge *Eureka*, during her maiden voyage up the Murray River in 1853

crossed the river in this crude craft with their horses swimming alongside. But the Aborigines evaded capture by taking to the water again. Eventually the eight canoes were found abandoned while the Aborigines had vanished into the scrub. (Bull, 1884, pp. 225-228).

Europeans had little more success in handling canoes than they had in making them. A number of men drowned in the Murray because they could not match the dexterity of the Aborigines in balancing these tiny craft. Le Souef (undated MS., p. 14) quotes several instances of his narrow escapes while using bark canoes: "The natives were splendid swimmers and soon taught me the art, and very useful it proved, as I should in the early course of my bush life have run the risk of being drowned scores of times, had I not been a strong swimmer. Shortly after I had learnt first to swim I was crossing the Goulburn one winter's day in a canoe by myself, when the river was in high flood. I was expert in the use of the bark canoes, which were made of a single sheet of bark, cut from a tree with a bend in it. The canoe I used that day was an old one, and the stern was level with the water, which was only kept out by a small bank of clay. I had got nearly three parts of the way across

The paddle steamer *Ellen*, seen here on the Queen's Reach, Chowilla, South Australia, in 1910, was typical of the river craft which replaced the bark canoes

the stream, when I noticed that the water was coming in fast, and looking back, saw to my dismay that the strong current had partially washed away the clay bank in the stern. A moment after, my frail craft sank from under me, leaving me struggling in the water in a heavy monkey jacket and strong boots. However I managed to swim into the head of a submerged tree and from thence to the bank."

On another occasion, a canoe being used by Le Souef and a companion to make a night crossing struck a submerged snag and split apart. They fell out in mid-stream, and had narrow escapes from drowning.

Such experiences sometimes had fatal conclusions. A memorial in Adelaide's West Terrace cemetery recalls one of the better-known fatalities. It was reported by Inspector Alexander Tolmer, who commanded the first Gold Escort that carried gold won by South Australian diggers on the Victorian goldfields to the Treasury in Adelaide. At one time he was Chief of Police for South Australia. The victims were two of his mounted constables; twenty-two year old John Dunning Carter, and twenty-four year old William Wickham. Tolmer had sent them to Moorundie, Edward John Eyre's station near

Blanchetown, to discourage Aboriginal attacks on overlanders bringing stock to Adelaide. (Tolmer, 1882, Vol. 1, pp. 239-240).

He wrote: "They had both gone up the Murray on duty, and when opposite Mr Wigley's station (six miles [9.65 km] from Lake Bonney) on the east bank of the river, they hobbled their horses and got into a native bark canoe with the intention of passing the night at the station, but in crossing both were drowned. It appeared to eye-witnesses who saw the fatal accident, that poor Carter was standing upright propelling the frail boat, whilst Wickham was sitting behind him in the stern, and upon leaning slightly over, upset it. As proof of the great

One of the last bark canoes to be seen at Murray Bridge, South Australia in 1910

The pressure of new wood, growing over the scar left by bark removal, has forced out the decaying core of this interesting canoe tree near the Finniss River in South Australia

This photograph, taken about 1875, shows the last survivors of the once powerful Morgan (South Australia) group of Aborigines

esteem in which they were both held by their comrades in the force, they subscribed a sum sufficient to remove their bodies into town [Adelaide], and erect a handsome monument to their memory in the West Terrace cemetery."

In 1947 the South Australian Police Force marked the centenary of the tragedy by renovating the grave and affixing a memorial tablet to the old grey slate headstone.

As settlement extended along the fertile river valleys, the contagious diseases

Bark for a canoe was cut from this red gum near the Finniss River, South Australia. The canoe would have been floated down the river for fishing on the lakes

The bark canoe cut from this red gum in the Ashbourne district of South Australia had to be carried eleven kilometres to Lake Alexandrina introduced by the settlers caused a rapid diminution of the Aboriginal population. Taplin (1874, p. 7) estimated that of the 3,000 members of the Narrinyeri group in 1840 only 600 of all ages had survived until 1873. This rapid decline was indicative of every district along the Murray. It was not long before bark canoes became a rare sight on the river.

On 9 December 1843, Hawker purchased a boat he had seen at Port Adelaide. The price, including sails, two oars, the rudder, and delivery to Moorundie, was twelve pounds. This was the beginning of the end for the bark canoes, although with their shallow draft they continued to be useful particularly on lagoons.

Ten years later, during August, September, and October 1853, Captain Randall in the *Mary Ann* and Captain Cadell in the *Lady Augusta* with her barge *Eureka*, demonstrated the potential of the Murray for commercial navigation. Their voyages ushered in the age of the paddle steamer, and were the prelude to a regular traffic of steamers serving an increasing number of settlers along the Murray Valley. It was not long before the Aboriginal and his bark canoe became a very rare sight. Ironically, the period of the river steamers was to be but a brief episode compared with that of the bark canoe, which had plied the river for many centuries.

The detailed record of the maiden voyage of the *Lady Augusta*, kept by two of the passengers, Messrs E. W. Andrews and James Allen, junior, from Goolwa to a point some 130 kilometres above Swan Hill, provides some interesting aspects: "Just before breakfast we saw some native women on the shore with a canoe. We hailed them, and one of their number held up a large fish, as if in token of amity. On our addressing them in friendly style, and inviting them to come alongside, several men made their appearance. This, I understand, is the usual custom of the natives when they apprehend danger, as they probably did in this instance, from the strange appearance of the steamer. The men conceal themselves until the women have established amicable relations with the strangers. On our repeated invitation a man and woman entered the canoe and came alongside. It was a mere piece of bark, about twelve feet [3.6 m] long and a yard [.9 m] broad, very slightly concave, and pointed at the bows. It had a little mud wall, forming a high gunwale all round, to keep out the water—the most primitive exhibition of naval architecture I ever saw or heard of. They both stood up in it, the man dexterously paddling it along with a pole about four yards [3.6 m] in length. This he held by the middle, working with either end as a stroke on one side or the other as required, while the woman balanced herself very skilfully, so as to keep their little bark in trim. They had three golden perch, exactly resembling the English fish in shape, and having a dorsal fin, but larger, and of yellowish tinge . . . they

One of the canoes issued by the Government of South Australia for use by Aborigines on the lower Murray River and on Lakes Alexandrina and Albert

let us have the whole, and we gave them in return a tin of flour." (Andrews, 1854, p. 204).

There can be little doubt that the sight of the first paddle steamers plying up and down the Murray River excited the curiosity of the Aborigines. Allen (1853, p. 21), one of the party on the *Lady Augusta*, tells something of their reactions: "Shortly after nightfall, and just as we were expecting to reach Chapman's, a native canoe came paddling out to us from the bank of the river, containing two or three natives, anxious to inspect the 'big one mongoe' (the native name for canoe), which had arrived in their waters. One of the natives remained on board, and piloted us up to Chapman's Station, which was a few miles further on, and opposite to which we anchored for the night at about 8 o'clock p.m. On our arrival there, the right-hand bank of the river was lined with natives, who, as a means of welcoming us, I suppose, had lit large bonfires along the edge of the river, which, with their dark figures flitting here and there before the flames, gave the scene a singularly wild appearance."

This canoe tree at Mannum, South Australia, is one of the few surviving trees with toe holds cut by the Aborigines to gain access to a suitable piece of bark for a canoe

The dead wood left by stripping bark for a canoe makes canoe trees highly susceptible to fires. This one is at Chowilla, South Australia

When the *Lady Augusta* reached the flat plains above Swan Hill, in September 1853, Allen (1853, p. 41) records how: "The dark figures of the natives, paddling their canoes through the reeds, fishing, has a peculiarly strange appearance, their heads only being visible through the reeds, and floating along without any perceptible support."

"The natives assembled here on our arrival in great numbers, and the ludicrous way in which they expressed their astonishment at the steamer was most amusing."

Three Aborigines had boarded the *Lady Augusta* at Wellington, apparently well pleased with the prospect of a novel voyage along the great waterway that had been a source of life to untold generations of their people. The long river journey by "the big one mongoe" must have been a profound experience although Allen (1853, p. 14) records that they accepted the situation almost without question: "They appeared perfectly at home in their new situation, and do not seem to express any surprise at the novelty of their position, so utterly uncontemplative are they. A gentleman on board asked one of them whether he could catch some fish for us. His reply was, that he could not then, as the steamer made 'big one run'."

The three Aborigines were jubilant upon their safe return and "displayed unmistakeable signs of joy on returning again to their own fatherland. Their relations and friends on shore, who, as we were told, believed they would never return after being absent so long, and who had absolutely been lamenting their death, came on board, when the congratulations and embracing which took place between them was most amusing." (Allen, 1853, p. 60).

While the white man's axe enabled expansion of the canoe-making industry, his territorial claims cut the Aboriginal off from his raw materials. This had drastic effects as the Aborigines could no longer fish in their traditional spots and obtain their own food. The Government could provide them with food for the body, but as in so many other cases they were robbed of sustenance for the spirit when they were barred from such traditional cultural pursuits as canoe-making and all that they meant to their pattern of life.

The South Australian Legislative Council concerned itself with the plight of the Aborigines in South Australia, including the populous lakes and lower Murray, as early as 1860. A Parliamentary Select Committee heard evidence from F. W. Howell, the Superintendent of Convicts, concerning Aboriginal problems. He testified that the land had been purchased around the lakes and that the owners had barred the Aborigines from cutting bark from gum trees. One of the greatest gifts the Government could give the people, he said, would be a canoe to each family. When the Committee suggested to Howell that canoes for the lakes could be obtained from as far as 160 kilometres up-river, where gum trees flourished, he thought it was preferable to provide the bark, thereby permitting the Aborigines to retain at least a part of their old occupation.

A sympathetic government was quick to respond, and canoes were made available. But five years later the Protector of Aborigines (Walker, 1865, p. 266) was reporting on the need for further assistance: ". . . the Wellington blacks are also very badly off for boats or canoes—those which were supplied to them by the Government, about six years ago, are all worn out. There are one or two bark canoes here which are very unsafe. One parted in the centre not long ago with two blacks in it, and one was very near being drowned in consequence. The blacks cannot now procure bark canoes as they could in former years. The want of boats is very much felt at Wellington where the tribes are divided, living on both sides of the river; and in bad weather I have known the sick, aged, and infirm to be almost starving for two or three days, because they had not a boat to come across to get their rations. The boats are also required for fishing and shooting—they cannot use their nets without the assistance of a boat, and they cannot cross the swamp or lagoons to shoot duck or swan except in a boat. The blacks on the Coorong are quite as bad if not worse off, as their living mainly depends on fishing and shooting; and they fish by nets there, and are seen to go into the water up to their armpits to set them, and the general consequence is that they get colds and die. New boats could be made for about £4 each, strong and serviceable. The blacks generally die from cold and privation, and very frequently their deaths may be attributed to their being obliged to go into the water in cold weather—being compelled by hunger to do so for want of a boat."

The last stronghold of the lower Murray people, known at the time as the Narrinyeri, was at Point McLeay where the Aborigines' Friends Association of Adelaide established a settlement in October 1859. The first missionary in charge was the Reverend George Taplin.

The people at Point McLeay, like those at Wellington and on the Coorong, were dependent upon spear fishing from their bark canoes for a livelihood and Taplin (1874, p. 30) noted that: "The men are very expert in the use of them."

A special correspondent, who visited Point McLeay in the mid-1870s, noted that the settlement took advantage of the Government supply of boats. He wrote: "The natives have an encampment on the shores of the lake, and near it are moored a number of strong roomy boats, in which they go out fishing or shooting. These boats are provided by the Government for the use of the blacks, and constitute their most valued possessions. It was a thoroughly sensible and practical idea to furnish the natives with these boats, because by their means they can obtain an independent living in a way closely associated with their former free mode of existence."

One of the last tenders called by the Government for canoes appeared in the *Mount Barker Courier*, of 23 and 30 September 1887. The tender of R. J. Blackwell, for the supply of twenty-six canoes for the Aborigines at the sum of £5.2.0 apiece, was accepted on 21 October.

In 1894 it was decided to establish a local boat building industry at Point McLeay to provide employment for the dwindling population and ensure the availability of suitable boats. A visitor to the settlement describes the newly established enterprise: "Down in a big woolshed on the lake shore we found Edward Kropingrie and Bertie Tripp (half-castes), and Albert Karloan (full-blooded native) engaged upon a pram 15 ft. [4.57 m] long, clinker built, well and securely ribbed and fastened with copper rivets, carefully if somewhat too strongly constructed from stem to stern. It was the first they had built, and it did them infinite credit.

"The light-draught flat-bottomed boats and the round-bottomed flat-bowed pram, which requires some practise to manage as it wobbles, like a boomerang, are much used by the natives for fishing and shooting on the lakes, and a boat-building industry is not a bad idea."

By June 1895 seven boats had been completed and these were distributed mainly to those Aborigines living away from the Mission and earning their own living by fishing. (Hamilton, 1895).

Conventional pram dinghies were a standard type of government work boat until World War II and were used increasingly by the Aborigines for their nomadic river journeys.

Today, those white men who still remember the Aborigines on the river can recall only these larger boats. However, Aboriginal oarsmen scorned the conventional thwart seat. They sat in the stern of the boat on an upturned box, facing the direction of travel, and pushing, not pulling, on the oars.

The bark canoe vanished from the scene in the early 1900s. One of the last examples is said to have had an inglorious ending as a feed box in a Morgan stable. Of the many hundreds of canoes used on the waters of the Murray-Darling basin, in western Victoria and the south east of South Australia, only two survive. One example, about three metres long and nearly a metre wide, was presented to the National Museum of Victoria in 1924 (Registered No. 29907). It was used at Bolton in the Swan Hill Shire of Victoria. The other canoe, a superb craft five metres long and nearly one metre wide, was made in 1904 at Avoca Station and used on the Darling River. It was presented to the South Australian Museum by D. H. Cudmore (Registered No. A6443).

Few Aborigines survive in the Murray-Darling basin today and in most districts only the scarred canoe trees remain as reminders of the canoe-making tradition carried on over many thousands of years.

In South Australia the canoe tree is synonymous with the Murray Valley. But it occurs also in even the deep hills—at Tungkillo, for instance, thirty kilometres or so from the Murray. The red gum is one of the most widely diffused

A decaying canoe tree being removed from the bank of the Murray River at Mannum, South Australia, to be taken to the South Australian Museum for preservation

Military barracks at Moorundie, on the Murray River, South Australia, about 1845

species of the eucalypts, and is found in every State except Tasmania. It prefers a moist fertile soil, so is found most often on the banks of streams and on alluvial flats. The species can reach thirty metres in height, and it is estimated that specimens have survived for 1,000 years. The River Murray banks offer ideal conditions, and the trees spread inland along the streams that fall from the hills to the main channel.

One might expect that the lake system near the mouth of the Murray would be fringed with red gums, but, until salt water was prevented from entering the Murray by the artificial barrages erected during the 1930s, the lakes were too saline to favour the growth of such trees. Upstream from the lakes the first red gums are seen some sixty kilometres from the mouth, a few kilometres below Murray Bridge, while on the tributary Finniss River the red gums begin about eleven kilometres from its junction with the Goolwa. These occurrences mark the limits to which the salt water used to penetrate from the sea at high tides or periods of low river.

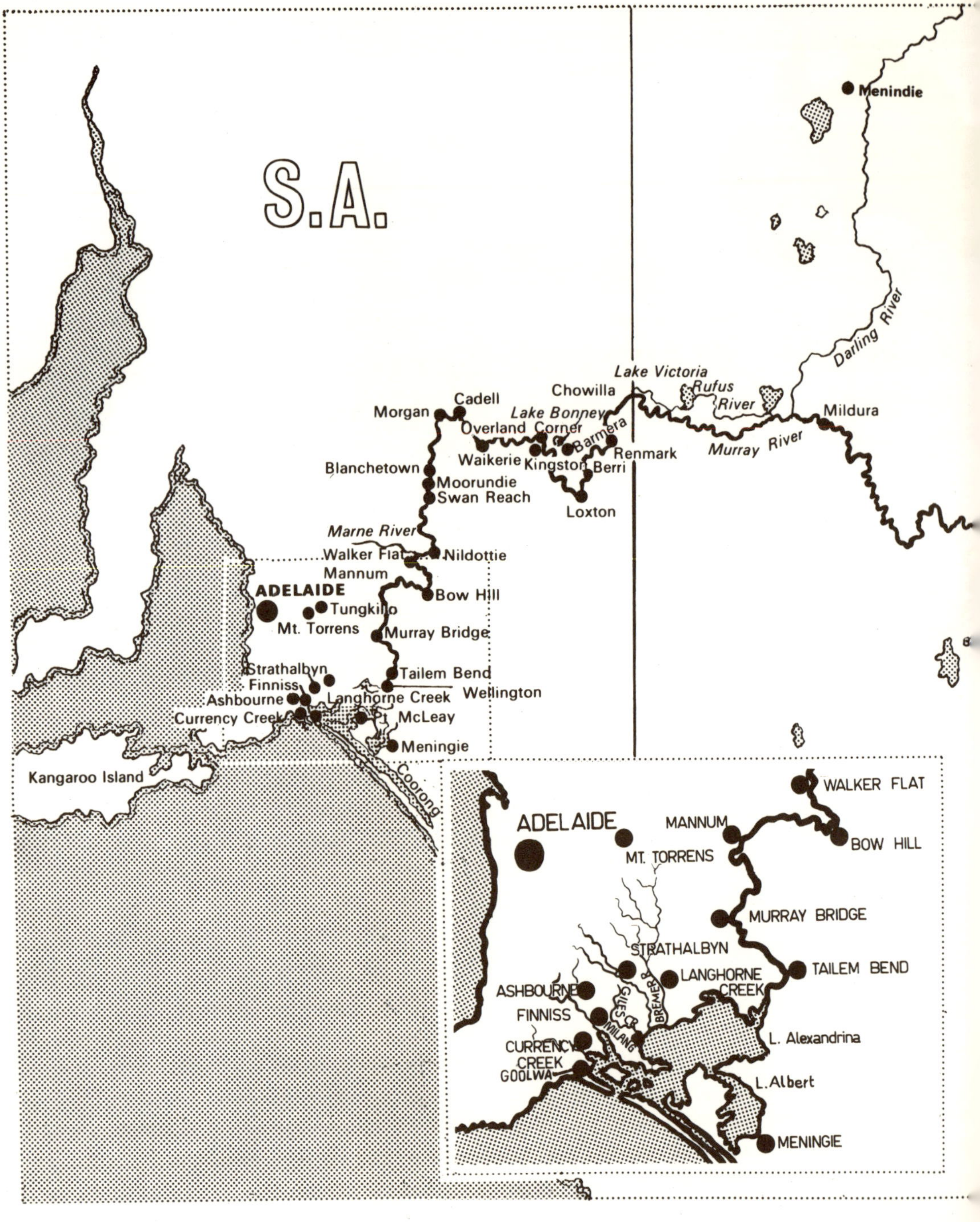
S.A.
Menindie
Darling River
Lake Victoria
Rufus River
Chowilla
Mildura
Murray River
Cadell
Morgan
Lake Bonney
Overland Corner
Barmera
Renmark
Waikerie
Kingston
Berri
Loxton
Blanchetown
Moorundie
Swan Reach
Marne River
Walker Flat
Nildottie
Mannum
ADELAIDE
Tungkillo
Bow Hill
Mt. Torrens
Murray Bridge
Strathalbyn
Tailem Bend
Finniss
Wellington
Ashbourne
Langhorne Creek
Currency Creek
McLeay
Meningie
Coorong
Kangaroo Island
WALKER FLAT
ADELAIDE
MANNUM
BOW HILL
MT. TORRENS
MURRAY BRIDGE
STRATHALBYN
TAILEM BEND
ASHBOURNE
LANGHORNE CREEK
BREMER R
FINNISS
MILANG
L. Alexandrina
CURRENCY CREEK
GOOLWA
L. Albert
MENINGIE

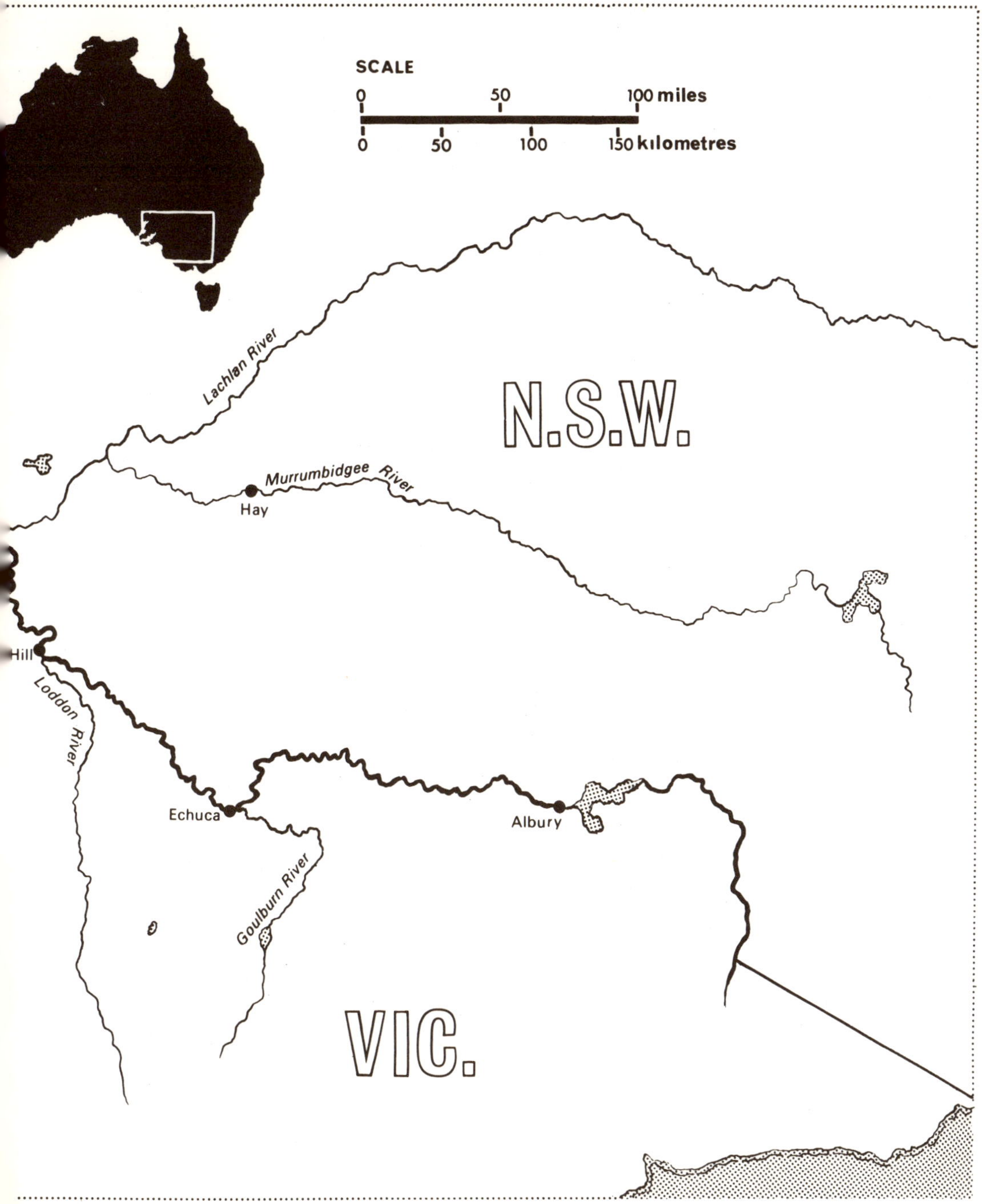
SCALE
0
50
100 miles
0
50
100
150 kilometres
N.S.W.
Lachlan River
Murrumbidgee River
Hay
Hill
Loddon River
Echuca
Albury
Goulburn River
VIC.

Soon after settlement the bark canoe gave way to dinghies and sail boats. Reproduced from a sketch by A. C. Kelly, a view on the lower Murray River about 1843

Several other streams flow from the Mount Lofty Ranges into the Murray River or Lake Alexandrina. Currency and Giles Creeks, small streams with sparse scatters of gums, have only a few canoe trees along their banks, but the Bremer, a larger river, has many fine canoe trees on its flats. Some of these trees date to the contact period as Taplin (1859, p. 66) records how the Aborigines went from Point McLeay to cut canoes on the Bremer at Langhornes Creek. This whole region numbers so many canoe trees that an indigenous boat-building industry must have flourished here to supply people living in treeless areas around the lakes and parts of the lower Murray River. It is interesting that local farmers can remember stories, passed on by earlier settlers, of canoes cut by the Aborigines and carried in bullock wagons over the thirty-two kilometres to Point Sturt on Lake Alexandrina. Many canoes may have been carried the long distance by Aborigines but during spring when the rocky, snag-bound hill streams became flooded the Aborigines would undertake the precarious task of floating them down to the lakes. (Black, 1947, pp. 351-361).

The South Australian Museum survey of canoe trees along the Murray River and its tributaries, and streams running into Lake Alexandrina, has been the means of locating several hundred examples. Their position has been plotted on maps, measurements taken of both tree and canoe scar size, and a photographic record made. Through these studies it is hoped to establish the extent and period of canoe-making by the study of regrowth over the scars left by bark removal. This varies considerably and on some trees the scars have almost completely healed. Dating methods are being investigated by the Radiocarbon Laboratory at the Australian National University, Canberra, to provide an accurate means of determining the age of canoe cuts.

Canoe-making is a lost art in southern Australia. It is essential that surviving canoe trees should be preserved as a memorial to the Aboriginal people and as irreplaceable relics of Australian history before the white man came.

BIBLIOGRAPHY

Allen, James, 1853: *Journal of an experimental trip by the 'Lady Augusta' on the River Murray*. Adelaide.

Andrews, E. W., 1854: Navigation of the Murray, *South Australian Almanack*, Adelaide.

Angas, George French, 1846: *South Australia Illustrated*. London.

Angas, George French, 1847: *Savage life and scenes in Australia and New Zealand*. 2 vols., London.

Berndt, Ronald M., 1941: The bark canoe of the Lower River Murray, South Australia. *Mankind, 3.*

Berndt, Ronald M. & Catherine H., 1964: *The world of the first Australians*. Sydney.

Beveridge, Peter, 1889: *The Aborigines of Victoria and Riverina*. Melbourne.

Black, E. Couper, 1947: The canoes and canoe trees of Australia. *Mankind, 3.*

Bull, John Wrathall, 1884: *Early experiences of life in South Australia*. Adelaide.

Curr, E. M., 1883: *Recollections of squatting in Victoria*. Melbourne.

Davidson, D. S., 1935: The chronology of Australian watercraft. *Journal of the Polynesian Society*, 44.

Eyre, E. J., 1845: *Journal of expeditions of discovery into Central Australia*. 2 vols., London.

Flinders, Matthew, 1814: *A voyage to Terra Australis*. 2 vols. and atlas, London.

Gerstaecker, F., 1853: *Narrative of a journey round the world*. 3 vols., London.

Hamilton, G., 1895: *Letter book of the Protector of Aborigines*. S.A. Archives.

Hawker, James C., 1841-45: *Diary of James C. Hawker*. S.A. Archives.

Hawker, James C., 1901: Early experiences in South Australia. *South Australian Register*. Adelaide.

Hume, Hamilton, 1831: *Journey of discovery to Port Phillip, New South Wales by Messrs. W. H. Hovell and Hamilton Hume in 1824 and 1825*. Sydney.

Hume, Hamilton, 1855: *A brief statement of facts in connexion with an overland expedition from Lake George to Port Phillip, in 1824*. Sydney.

Le Souef, A. A. C., undated: *Personal recollections of early Victoria.* Typewritten MS. South Australian Museum.

McCarthy, F. D., 1944: The making of a Bark Canoe. *Aust. Mus. Mag.* 8 (6).

Mitchell, Major T. L., 1838: *Three expeditions into the interior of Eastern Australia.* 2 vols., London.

Oxley, John, 1820: *Journals of two expeditions into the interior of New South Wales.* London.

Parliamentary Paper, 1860: Report of the Select Committee of the Legislative Council upon "The Aborigines" 165, Adelaide.

Roth, Walter E., 1910: North Queensland ethnography. Bulletin No. 14. *Rec. Aus. Mus.*, 8 (1).

Smith, W. Ramsay, 1930: *Myths & legends of the Australian Aboriginals.* London.

Smyth, R. Brough, 1876: *Geological survey of Victoria.* Melbourne.

Smyth, R. Brough, 1878: *The Aborigines of Victoria.* 2 vols., London.

Spencer, Baldwin & Gillen, F. J., 1904: *The northern tribes of Central Australia.* London.

Sturt, Captain Charles, 1834: *Two expeditions into the interior of Southern Australia.* 2 vols., London.

Sturt, Captain Charles, 1849: *Narrative of an expedition into Central Australia.* 2 vols., London.

Muller, Frederik, 1898: *Abel Janszoon Tasman's Journal.* 1642, Ed. Frederik Muller & Co., Amsterdam.

Taplin, Rev. George, 1859-79: *Journal of the Rev. Geo. Taplin.* S.A. Archives.

Taplin, Rev. George, 1874: *The Narrinyeri: an account of the tribes of South Australian Aborigines.* Adelaide.

Tindale, Norman B., 1926: Natives of Groote Eylandt and of the west coast of the Gulf of Carpentaria. *Rec: S. Aust. Mus.*, 3 (2).

Tindale, Norman B., 1935: The legend of Waijungari, Jaralde Tribe, Lake Alexandrina, South Australia, and the Phonetic system employed in its transcription. *Rec. S. Aus. Mus.*, 5 (3).

Tolmer, Alexander, 1882: *Reminiscences of an adventurous and chequered career at home and at the Antipodes.* London.

Walker, John, 1865: Quarterly report of Protector of Aborigines. *S.A. Gov. Gazette.*

ACKNOWLEDGMENTS

Preparation of this book has been facilitated by the generous assistance of many individuals and institutions. Particular thanks are due to Professor John Mulvaney, of the Australian National University, Canberra, for his continuous encouragement and advice. Information and photographs were supplied by Marjorie Angas, Doug Bannear, Joy Chilman, Robert Edson, Harry Godson, Richard Kimber, John Ledo, Henry Lucas, Bill Reschke, Jean Schmaal, R. G. Schubert, and Mr Alan West, the Curator in Anthropology at the National Museum of Victoria. Mr Denis Marfleet made the photographic enlargements for reproduction, members of the staff of the Archives Department of the State Library of South Australia gave generously of their time, and many landowners in the Murray Valley willingly assisted in the survey of canoe trees carried out by the South Australian Museum.

ILLUSTRATION SOURCES

Photographs by the author: In colour: Canoe tree at Nor-West Bend Station; healed scar on canoe tree; two canoe trees at Nor-West Bend Station; canoe tree at Chowilla. In black and white: Frontispiece, canoe tree at Blanchetown; p. 19, pleasure craft on Murray River; p. 23, canoe tree near Giles Creek; p. 32, canoe tree at Milang; p. 38, canoe tree at Blanchetown; p. 40, canoe tree at Chowilla; p. 43, canoe tree near Hindmarsh River; p. 51, canoe tree at Blanchetown; p. 56, canoe tree with decaying core; p. 58, canoe tree near Finniss River; p. 59, majestic canoe tree at Ashbourne; p. 62, canoe tree with toe holds and canoe tree damaged by fire; p. 66, canoe tree being removed to South Australian Museum. *South Australian Museum:* p. 17, Murray River Aborigines; p. 21, bullock team crossing Murray River in drought; p. 34, Aboriginal and child in canoe; p. 34, bark canoe on Wimmera River (photo by Mr R. Bell); p. 49, bark canoe made at Avoca; p. 50, photograph of canoe tree taken eighty years ago; p. 55, Bark canoe at Murray Bridge. *Illustrations from the Taplin Collection in the South Australian Museum:* p. 16, group of bark canoes; p. 25, "Pelican"; p. 27, Teen-minne; p. 44, Ngulgare; p. 45, Narappane; p. 46, Moilge; p. 48, Toldinyeri; p. 61, canoe issued to Aborigines by the Government. *State Library of South Australia* (Archives Department): p. 9, floating wurlies; p. 28, bark canoe being made (photo by George Burnell); p. 53, the *Lady Augusta* (sketch by J. H. Adamson); p. 67, barracks at Moorundie (watercolour presented by Miss L. M. Eyre); p. 70, bark canoe and sail boat on the Murray River. *State Library of South Australia* (Godson collection); p. 39, canoe trees at Waikerie; p. 34, paddle steamer *Ellen;* p. 41, Aborigines fishing from canoe at Chowilla; p. 57, survivors of the Morgan Aborigines. *South Australian Lands Department:* p. 20, low river at Waikerie. *Mr D. Darian Smith:* p. 19, Murray River in flood at Renmark. *Mr C. P. Mountford:* p. 11, bark painting of Malay Prau. *The Natural History of Man:* p. 31, engraving of fishing canoes. *Royal Geographical Society of Australasia* (*South Australian Branch*): p. 36, "River Murray Dan". *Two Expeditions into the Interior of Southern Australia* (Captain Charles Sturt): p. 8, sketch of junction of Murray-Darling Rivers. *National Museum of Victoria:* p. 11, canoes on Lake Tyers (Photo by Mr N. C. Caire). *Illustrated Melbourne Post:* p. 29, funeral procession of canoes. *Australian Institute of Aboriginal Studies:* p. 12, dug-out canoe under construction (photo by Nicolas Peterson). *Trustees of the British Museum of Natural History:* Aborigines and bark canoes at Port Jackson (by "Port Jackson Painter"). *South Australia Illustrated:* endpapers (Aboriginal artefacts); p. 15, Murray River above Moorundie by G. F. Angas. *Journals of Expeditions of Discovery into Central Australia* (by Edward John Eyre): p. 15, Government post at Moorundie (sketch by G. Hamilton). p. 8, tied bark canoes at Port Jackson. Photo from *Voyage de Découvertes Aux Terres Australes, M. F. Peron* (*Atlas par Mm. Lesueur et petit*):

NOTE: All page numbers quoted above refer to those in this volume.

INDEX

Aborigines' Friends Assoc., 64
Albury, 28
Angas, George Fife, 17
Angas, George French, 17
Arnhem Land, 10
Avoca Station, 65

Barge, *Eureka*, 37, 60
Bark painting, 10
Berndt, Prof. R., 22
Beveridge, P., 20
Black, Dr E. Couper, 7
Blackwell, R. J., 64
Blanchetown, 18, 55
Boat building, by Aborigines, 65
Bremer, the 70
Bugis, 10

Cadell, Capt., 60
Canoes, bark:
accidents, 37, 53, 54
manufacture by Aborigines, 7, 14, 16, 17, 18, 29, 30, 31, 41, 42, 60
manufacture by Europeans, 16, 30
use by Aborigines, 20, 35, 52, 60, 63
use by Europeans, 17, 21, 29, 42
use by explorers, 31, 42
use by geologist, 52
use by pastoralists, 42, 47, 52
sewn, manufacture by Aborigines, 9, 10
tied, manufacture by Aborigines, 7
tied, manufacture by Europeans, 16
tied, use of by Europeans, 16
Canoes, dug-out, 9, 10
Canoes, Government supply of, 64, 65
Canoes, log, manufacture by Europeans, 28
Cape York Peninsula, 14
Cave art, 10
Celebes, 10
Ceremonies, 22, 42
Coorong, 26, 64
Cumpungo, 21
Curr, E. M., 42
Currency Creek, 70

Darling River, 24, 30, 37, 39, 42, 65

Eureka, barge, 37, 60
Eyre, Edward John, 17

Finniss River, 67
Fishing, Aborigines, 18, 20, 22
Flinders, Matthew, 14

Gawler, Governor George, 21
Gerstaeker, F., 28, 29, 42
Gifts, exchange of, 22
Giles Creek, 70
Goldfields, Victorian, 42
Goolwa, 60, 67
Goulburn River, 42, 53

Hawker, J. C., 20, 29, 30, 31, 60
Hovell and Hume, 30
Howitt, A. W., 52
Hume, Hamilton, 16, 30, 42

Indonesians, 14

Kangaroo Island, 26
Karloan, Albert, 24, 65
Kropingrie, Edward, 65

Lachlan River, 42
Lady Augusta, 37, 60, 63
Lake Albert, 22
Lake Alexandrina, 22, 70, 71
Lake Bonney, 55
Langhornes Creek, 70
Legislative Council, South Australian, 63
Le Souef, A. A. C., 30, 35, 52, 53, 54

Macarthur River, 10
Macassan influence, 10
Mary Ann, 60
Mildura Station, 52
Mitchell, Major T. L., 42
Mitchell River, 52
Moorundie, 18, 21, 29, 35, 54, 60
Mount Misery, 24
Murray Bridge, 67
Murrumbidgee River, 16, 30
Myth, Ngurunderi, 22, 24

Narrinyeri, 64
New Guinea, 14
New South Wales, 7, 31
Northern Australia, 10

Outrigger, 10
Oxley, John, 14

Paddle steamers, 7, 37, 60, 63
Paddle steamer *Lady Augusta*, 37, 60, 63
Paddle steamer *Mary Ann*, 60
Parliamentary Select Committee, 63
Pellew Islands, 10
Pestonjee Bomanjee, 21
Piomubko, 21
Point McLeay, 24, 64, 65, 70

Point Sturt, 70
Police Force, South Australian, 57

Queensland, 9, 14

Randall, Capt., 60
Rufus River, 52

South Australian Legislative Council, 63
South Australian Police Force, 57
Spencer, Sir Baldwin, 10
Sturt, Capt. Charles, 14, 30, 39
Swan Hill, 14, 20, 30, 37, 60, 63, 65

Taplin, Rev. George, 22, 60, 64, 70
Tasman, Abel, 7
Tindale, Dr N. B., 24
Tolmer, Inspector A., 54
Torres Strait, 14
Trepang fishermen, 10
Tungkillo, 65
Turtle catching, by Aborigines, 21

Victoria, 7, 31, 65

Wellington, 33, 63, 64

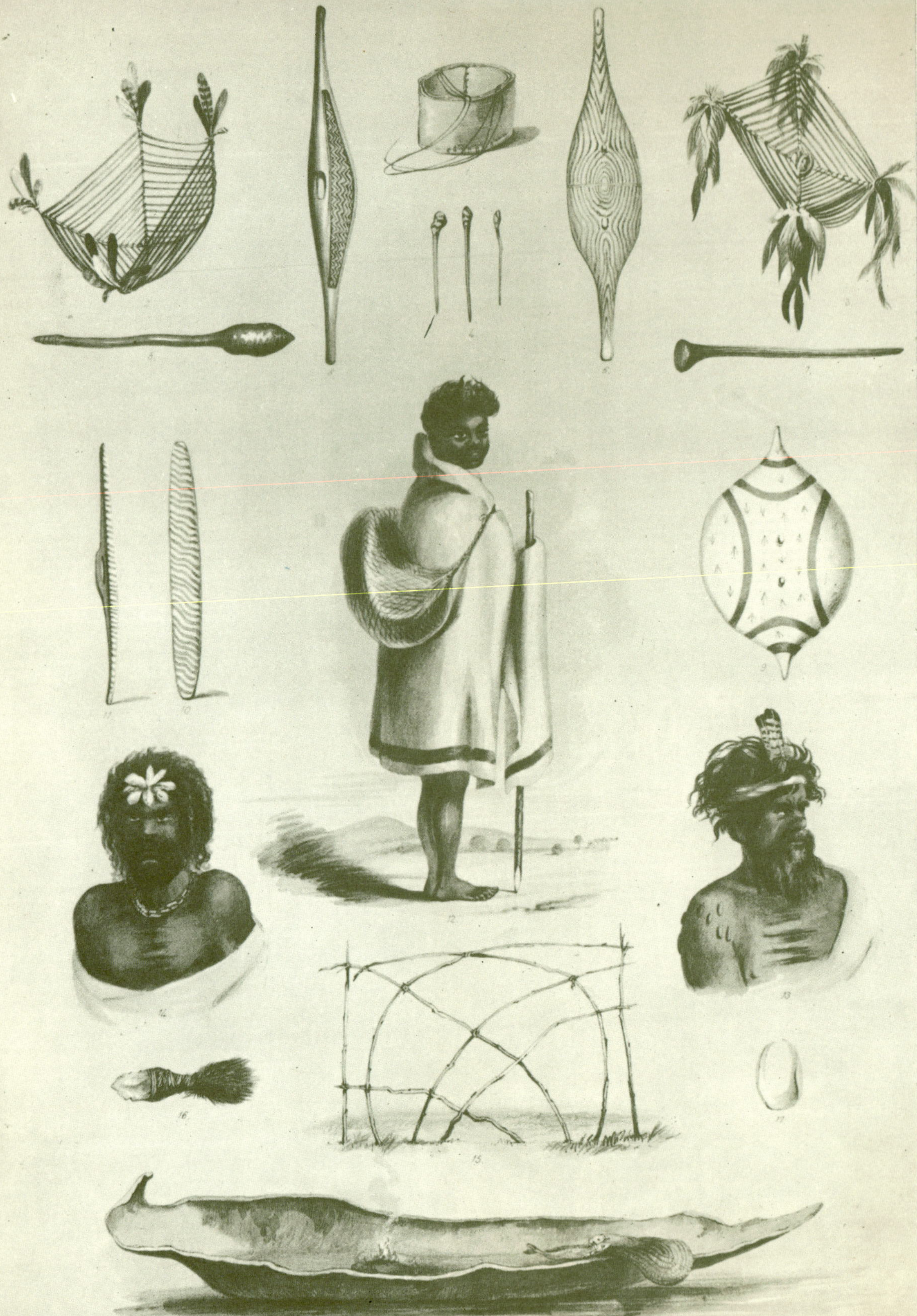

11
10
9
12
14
13
16
15
17
18

JANUSZ PIEKALKIEWICZ

DER VW KÜBELWAGEN TYP 82 im Zweiten Weltkrieg

MOTORBUCH VERLAG STUTTGART

Umschlagzeichnung: Carlo Demand
Umschlagkonzeption: Siegfried Horn

ISBN 3-87943-468-9
2. Auflage 1984

Eine Abteilung des Buch- und Verlagshauses Paul Pietsch GmbH & Co KG

Satz und Druck: Schwabenverlag, 7302 Ostfildern 1 (Ruit)
Bindung: Großbuchbinderei Ernst Riethmüller & Co., 7000 Stuttgart
Printed in Germany

INHALTSVERZEICHNIS

Vorwort	7
Der VW-Kübelwagen Typ 82 im 2. Weltkrieg	9
Daten + Fakten	160
Bibliographie	190
Danksagung	191
Bildquellen	191

VORWORT

»Der Wagen soll der Freude des deutschen Volkes dienen...« versicherte am Himmelfahrtstag, dem 26. Mai 1938, Adolf Hitler bei der Grundsteinlegung zum Volkswagenwerk in Fallersleben, dem Geburtsort des »Deutschland – über – alles« Dichters Hoffmann. Das störte den Führer keinesfalls, schon zwei Tage später den Befehl zur Vorbereitung des bewaffneten Einmarschs in die Tschechoslowakei zu geben.

Während der kleine Mann auf der Straße immer noch hoffte, daß der Volkswagen »ihm vor allem an Sonn- und Feiertagen zur Quelle eines unbekannten, freudigen Glücks würde« (Adolf Hitler), hatte das Heeres-Waffen-Amt dem Konstrukteur Porsche über die Schulter geschaut – und nicht ohne Wissen des Führers – seine Forderungen nach einem »leichten Einheits-PKW« für die Wehrmacht geltend gemacht. Das war die Geburtsstunde des berühmten VW-Kübelwagens. Und mit Ausbruch des Krieges wird schnell von Volksmotorisierung auf Truppen-Motorisierung umgeschaltet.

Das flinke, zähe Gefährt wird bald genauso von Freund wie Feind geschätzt. Und selbst die US-Streitmacht, der man keinen Mangel an Kraftfahrzeugen nachsagen kann, hat noch rechtzeitig am Tage der Invasion auf den Kontinent – am 6. Juni 1944 – eine reichhaltige Broschüre »War Department Technical Manuel: *German Volkswagen*« den eigenen Truppen ausgehändigt, um die Benutzung der Beutewagen zu erleichtern. Das kam nicht von ungefähr, der VW 82 hatte bereits seine Probe auf allen Kriegsschauplätzen gut bestanden. Auch die harten russischen Winter konnten ihm nichts ausmachen, und dort, wo die Räder der schweren Kübelwagen von Horch und Tatra im grundlosen Morast der Mittelfront oder im Sand ausgedörrter Sonnenblumenfelder durchdrehten, fraß sich der leichte Volkswagen vorwärts. Drei Mann konnten zur Not einen Volkswagen anheben, während den schweren Tatra sogar ein Halbzug nicht zu bewegen vermochte.

Selbst Feldmarschall Rommel bedankte sich eines Tages persönlich bei Porsche: »Sie haben mir das Leben gerettet.« Als Porsche ihn erstaunt ansah, weil er nicht wußte, was gemeint war, hörte er weiter: »Ihr VW-Kübelwagen, den ich in Afrika benütze, löste beim Überfahren eines Minenfeldes nicht aus, während der mir folgende schwere Horch mit dem Gepäck hochging.«

So mancher Leser wird beim Betrachten der Fotos sich in die damalige Zeit versetzt fühlen

und wieder an die eigenen Erlebnisse mit einem VW 82 erinnert.
Gewidmet ist dieser Band allen denen, die Interesse an Kraftfahrzeugen des Zweiten Weltkrieges und besonders an diesem unverwüstlichen Kübelwagen haben.

(Janusz Piekalkiewicz)

1

Es ist Mitte Januar 1931 – die Weltwirtschaftskrise hat gerade ihren Höhepunkt erreicht – als die Chefs der deutschen Automobilwerke ein Schreiben mit recht umfangreichen Anlagen erhalten. Der Absender, ein österreichischer Konstrukteur namens Dr. Ing. h. c. Ferdinand Porsche, teilt darin mit, daß er soeben ein eigenes Konstruktionsbüro in Stuttgart eröffnet hat:

DR. ING. H. C. FERDINAND PORSCHE
STUTTGART, 13. Jan. 1931
KRONENSTRASSE 24
TELEPHON 24301

Jch bin am 3. September 1875 in Maffersdorf bei Reichenberg geboren und begann meine Tätigkeit mit 18 Jahren bei den Vereinigten Elektrizitätswerken vorm. Egger & Co. in Wien. Nach vier Jahren war ich Leiter des Prüfungsraumes und Assistent des Berechnungsbüros dieser Firma.

Da ich mich frühzeitig schon für den Automobilbau interessierte, konstruierte ich bis zum Jahre 1900 mein erstes rein elektrisches Automobil und zwar mit Radnaben Elektromotor, den sog. Lohner-Porsche-Wagen. Dieses Fahrzeug war bereits auf der Weltausstellung 1900 in Paris.

Kurz darauf brachte ich den ersten Mixte-Wagen (Benzin-elektrisch getrieben) mit Vorderrad-Antrieb heraus. Hierfür erhielt ich im Jahre 1902 den Pötting Preis.

Jm Jahre 1906 trat ich als Direktor und Chefkonstrukteur bei Austro-Daimler in Wiener Neustadt ein, wo ich im Jahre 1917 General-Direktor wurde. U. a. baute ich dort Wagen für die Prinz Heinrich Fahrt im Jahre 1910, bei welcher es mir gelang, mit meinen Fahrzeugen den 1., 2. und 3. Preis zu erringen. Anschließend daran konstruierte ich Alpen-Fahrzeuge, die in diversen Jahren sich ebenso glänzend bewährten. Daß Hand in Hand mit dieser Entwicklung auch die Serien-Fahrzeuge entsprechend behandelt wurden, ist selbstverständlich.

Für die österr. Militärbehörde wurden von mir folgende Konstruktionen entwickelt: Zugwagen, Seilwindewagen, Landwehrtrain-Fahrzeuge mit 5 Anhängern, 30,5 cm Mörser Zugwagen, diverse Flugmotoren, von denen in England u. a. Lizenzen genommen wurden und tausende auf Grund dessen entwickelt, Motorisierung der 42 cm Mörser. Ferner baute ich für diese Firma Feuerwehrfahrzeuge, elektrische Oberleitungs-Omnibusse.

Jn meiner Nachkriegstätigkeit konstruierte ich für Austro-Daimler den 6 Cylinder A D Wagen, den 6 Cylinder A D V Wagen und den 6 Cylinder A D M Wagen.

Jm Jahre 1924 trat ich bei der Daimler A.-G. Untertürkheim als Direktor und Vorstandsmitglied ein. Für diese Firma entwickelte ich ca. 20 Personenwagen Typen, darunter die bekannten Kompressor-Typen wie z. B. SS und SSK Sport- bzw. Rennmaschinen. Ebenso entwickelte ich Rennwagen für die Targa Florio 1924 und für den großen Preis der Nationen. Sämtliche heute noch serienmäßig gelieferten Typen von Mercedes sind meine Konstruktion bis auf den 7,7 Ltr. Kompressor, dessen Grundtype ebenso von mir stammt.

Ferner entwickelte ich für diese Firma 14 Lastwagentypen, darunter einen großen Sechsradwagen; ebenso sind die bekannten Mercedes Rohöl-Lastwagen von mir konstruiert.

Weiterhin entwickelte ich in dieser Zeit 2 & 3 Cylinder Kleinflugzeug-Motoren wie auch einen 1000 PS wassergekühlten und 800 PS luftgekühlten Flugmotor.

Jm Jahre 1929 trat ich bei den Steyr-Werken in gleicher Eigenschaft wie bei Daimler-Benz als Direktor und Vorstandsmitglied ein und entwickelte für diese Firma einen 6 Cylinder 2 Ltr. Wagen, wie auch eine 8 Cylinder 5 Ltr. Type. Beide Typen waren bereits auf den Ausstellungen in Paris und London und sind hervorragend beurteilt worden.

Jch bin Jnhaber mehrerer hoher Orden, in Jtalien Cavaliere, Ehrendoktor der technischen Hochschule in Wien und Ehrendoktor der technischen Hochschule in Stuttgart.

Jch habe selbst an verschiedenen sportlichen Veranstaltungen teilgenommen und öfters Preise errungen.

Jch bin verheiratet, meiner Ehe sind 2 Kinder entsprossen, eine Tochter, Luise Piech geb. Porsche, geb. 29. 8. 1904 und ein Sohn, Ferdinand, geb. am 19. 9. 1909, letzterer ist als Konstrukteur bei mir beschäftigt.

Da mir nachgesagt wird, daß meine Arbeitskraft so groß sei, um mehrere Werke gleichzeitig mit Konstruktionen etc. zu speisen, habe ich nunmehr beschlossen, ein eigenes Konstruktionsbüro zur Lieferung von Automobil- & Motor-Konstruktionen wie auch zur Beratung und Unterstützung der einschlägigen Jndustrie zu eröffnen. Das Büro ist bereits mit Aufträgen versehen und befindet sich in Stuttgart, Kronenstraße 24.

gez. Ferdinand Porsche.

Am 25. April 1931 wird die »Dr.-Ing. h. c. F. Porsche GmbH, Stuttgart, Kronenstraße 24, Konstruktion und Beratung für Motore und Fahrzeuge« im Handelsregister eingetragen. »Es war schlecht, sehr schlecht«, erinnert sich heute *Ghislaine E. J.* Kaes, Porsches Neffe und Sekretär. »Wir lebten als Firma von der Hand in den Mund, sozusagen.«

Hier in der Garage seiner im Landhausstil erbauten Villa auf dem Killesberg, im Norden von Stuttgart, entwickelt Porsche jedoch Konstruktionen, die später zu den wichtigsten Merkmalen des Volkswagens gehören. So wird am 10. August 1931 die in der Kronenstraße konzipierte Drehstab-Federung als Patent angemeldet. Sie entsteht im Auftrag der Wanderer Werke.

Diese neue Radaufhängung, eine revolutionäre Erfindung, zählt zu den radikalsten Neuerungen des zukünftigen Volkswagens. Porsche hält nämlich die Blatt- und Schraubenfedern für rückständig. Hohes Gewicht und komplizierte Wartung lassen beide Federungssysteme für ein zugleich leichtes und leistungsfähiges Fahrzeug der Mittelklasse denkbar ungeeignet erscheinen. Und er löst das Problem durch die Drehstabfederung, auch Torsionsfederung genannt. Sie beruht auf einem Prinzip, wonach ein Stab, der an einem Ende fest lagert,

durch die Auf- und Abschwingungen eines Rades von seinem anderen Ende her verdreht wird. Der aus Stahlblattfedern gefertigte Stab verwindet sich also beim Abfangen von Unebenheiten. Je härter der Stoß, desto mehr verwindet und spannt er sich und setzt dabei weiterer Verwindung einen um so stärkeren Widerstand entgegen – ein Vorgang, der auch beim Biegen eines normalen Gartenschlauches aus Gummi erkennbar wird. Die Blattfedern sind in der Mitte fest eingespannt und zusammengeklammert, um ein Verdrehen zu vermeiden, und mit an den Vorderrädern des VW befestigten Traghebeln verbunden. Die ähnlich angeordnete Hinterradaufhängung verwendet für jedes Rad je einen Drehstab, der an seinem äußeren Ende gekerbt und verstellbar ist, um eine geometrisch einwandfreie Hinterradaufhängung zu gewährleisten. Jeder Stab ist aus Stahl und für links- oder rechtsseitige Verwendung vorgespannt.
Auch auf dem Killesberg entsteht für die Zündapp GmbH Nürnberg Porsches sechster Kleinwagen. Am 9. September 1931 beginnt er mit der Arbeit, und nach knapp drei Monaten, am 1. Dezember 1931, ist die Karosserie-Skizze fertig. Sie zeigt deutlich schon die Käferform, die eines Tages zum Sinnbild des Volkswagens wird.
Die Russen, deren Automobilbau noch in den Kinderschuhen steckt, laden Porsche 1932 nach Moskau ein. Er soll die Rote Autoindustrie auf die Beine stellen. Ein unbeschränktes Bankkonto, eine Prachtvilla auf der Krim und viele andere Extras stehen ihm bevor. Porsche besichtigt das wichtigste Zentrum der Industrie und sagt nach seiner Rückkehr: »Ich konnte mich in einer Gießerei mit den Arbeitern nicht verständigen und ohne Verständigung kann ich nicht wirken. Und in meinem Alter« – er war damals 57 – »lerne ich die Sprache nicht mehr.«
Im Jahre 1934 treffen sich im Kaiserhof in München zwei ehemalige Untertanen des k.u.k.: Porsche und Hitler. Sie führen ein fast freundschaftlich verlaufendes Gespräch über die Idee eines volkstümlichen Autos, dem auch Hitlers Freund Jakob Werlin beiwohnt, der als Münchner Mercedes-Händler dem Führer im Jahre 1923 den ersten Mercedes verkaufte. Werlin, der Hitlers Autobesessenheit kräftig zu schüren versteht, wird später mit dem 33. Erlaß des Führers vom 16. Januar 1942 zu seinem Generalinspektor für das Kraftwesen ernannt und hat bis zum bitteren Ende jederzeit unangemeldet Zutritt zum ›Allerhöchsten‹.
Hitler, von Porsches Vorschlägen recht angetan, steuert allerdings auch eigene Gedanken bei: Nach seiner Auffassung dürfe ein solcher Wagen keineswegs mehr als sieben Liter Kraftstoff auf hundert Kilometer verbrauchen – einen Liter weniger als Porsche erwägt – müsse einer vierköpfigen Familie Platz bieten, zum Schutz gegen das Einfrieren im Winter luftgekühlt sein und solle eine 100 km/h Geschwindigkeit nicht nur erreichen, sondern als Dauergeschwindigkeit einhalten können.
Übrigens hat man über die Verwendung eines Volksautos ziemlich früh genaue Vorstellungen. Schon am 11. April 1934 findet im Reichsverkehrsministerium eine Besprechung statt, an der auch unter anderem Vertreter der Reichskanzlei und des Reichspropaganda-Ministeriums teilnehmen. Man ist sich einig: Der Anschaffungspreis eines populären Vehikels darf höchstens 1000 Reichsmark betragen, die Betriebskosten sollen 6 Pfennig je Kilometer nicht übersteigen und in dem Wagen Sitzplätze für drei Erwachsene und ein Kind sowie genügend Raum für Gepäck sein. »Diese Bedingun-

gen entsprechen auch den militärischen Erfordernissen, da sich nach Entfernung des Aufbaus drei Mann, ein Maschinengewehr und Munition unterbringen lassen«, stellen die Herren in ihrem Abschlußbericht fest.
Adolf Hitler sieht Porsche offensichtlich gern. Es mag die Ausstrahlung des berühmten Konstrukteurs sein, der nicht in Uniform erscheint und seine österreichische Sprache nie verfälscht, der sich gibt, wie er ist, einfach und doch groß. Folgt Porsche einem Ruf Hitlers zum gemeinsamen Abendessen, pflegt ihn der Gastgeber mit den Worten zu empfangen: »Ich habe für Sie Würstel und Bier richten lassen, ist das recht?« Dann sitzen beide, mit Bleistift und Papier bewaffnet, und besprechen technische Einzelheiten einer Neukonstruktion. Später einmal, im Dezember 1941, als Porsche nach Orel an die russische Front geflogen werden soll, um die eigenen und die erbeuteten Panzer zu begutachten, befiehlt Hitler, seinem eigenen Piloten, Oberst und Flugkapitän Hans Baur, Porsche dorthin zu bringen. Denn, so erklärt er: »Generale habe ich genug, da kann einmal einer abstürzen, Porsche habe ich nur einen.«
Endlich am 27. April 1934 wird die Skizze Nr. 804, eine ›5-Sitzer-Limousine‹, die erste Skizze des Porsche Typ 60 – Volkswagen – vom Zeichenbrett genommen.
Am 22. Juni 1934, schließt Dr.-Ing. h. c. Ferdinand Porsche mit dem Reichsverband der Deutschen Automobilindustrie (RDA) einen Volkswagen-Konstruktionsvertrag, »um die Motorisierung des deutschen Volkes auf der Grundlage einer Gemeinschaftsarbeit unter Einsatz der besten Kräfte des Automobilwesens mit allen Mitteln zum Wohle des deutschen Reiches zu fördern«. Der RDA seinerseits denkt überhaupt nicht daran, sich auf ein Gemeinschaftsprojekt einzulassen, das seine eigenen Verkaufsziffern unweigerlich reduzieren würde. Er tut wenigstens so, als wäre er zur Mitarbeit bereit.
Bei der Eröffnung der Berliner Automobilausstellung im Jahre 1935 gibt sich Hitler, obwohl die Versuchsmodelle noch nicht fertig sind, optimistisch. »Ich freue mich«, spricht er, »daß es dank der Fähigkeit des glänzenden Konstrukteurs Porsche unter Mitarbeit seines Stabes gelungen ist, die Vorentwürfe für den deutschen Volkswagen so weit fertigzustellen, daß die ersten Exemplare Mitte dieses Jahres erprobt werden können. Es muß möglich sein, dem deutschen Volk einen Kraftwagen zu schenken, der im Preis nicht mehr kostet als früher ein mittleres Motorrad und dessen Benzinverbrauch mäßig ist!«
Als Hitler zu seiner Bestürzung hört, Porsche sei kein deutscher Staatsangehöriger, veranlaßt er sofort dessen »Germanisierung«. Am 17. Dezember 1934 wird Porsche mitgeteilt: »Der Führer wünscht, daß der größte deutsche Kraftwagenkonstrukteur die deutsche Staatsbürgerschaft erwirbt.«
Ghislaine Kaes: »Zu mir, der ich ihm das Schreiben vorgelegt hatte, sagte er wörtlich: ›Da wer'n ma ja nix machen können. Tust mir das erledigen.‹ Als Österreicher geboren, besaß er ab 1918 die tschechische Staatsbürgerschaft. Beim Tschechischen Konsulat in Stuttgart mußte Porsche die Ausbürgerungserklärung selbst unterschreiben. Diesen Weg hatte ich ihm nicht abnehmen können. Porsche bemühte sich um einen freundlichen Ton, der Konsul aber blieb eiskalt verschlossen. Als wir dann das Konsulat verließen und die Stufen des Gartenweges hinabschritten, hing Porsche noch seinen Gedanken nach. Da blieb er auf einmal stehen und sagte versonnen vor sich hin: ›Der hat uns aber nicht mögen.‹

Nun entschloß sich der Professor zu einer Reise in die Vereinigten Staaten, um die Massenproduktion von Automobilen an Ort und Stelle zu studieren. Am 19. September 1936 bat mich Porsche, seinen USA-Trip vorzubereiten. Nur er und ich sollten reisen. Wie er mir sagte, sei er angehalten ein Werk zu bauen in dem zuerst jährlich 500 000 Wagen, später bis 1,5 Millionen Volkswagen im Jahr hergestellt werden können. Weil es dafür in Europa keine Vorbilder gab, fuhren wir also los.«

Am 12. Oktober 1936 werden die drei VW-Prototypen, die sogenannten V 3 – übrigens bei Mercedes Benz in Untertürkheim hergestellt – dem Reichsverband der Deutschen Automobilindustrie (RDA), ganze 16 Monate nach dem vereinbarten Liefertermin, zur Erprobung übergeben. Professor Porsche selbst weilt gerade in den USA und besucht zur gleichen Stunde – man feiert das große amerikanische Fest, den Columbus-Tag – das erste Vanderbilt-Cup-Rennen.

Berlin, Automobilausstellung, 1938: Vor dem Chassis des KdF-Wagens (VW 60), links Dr. Robert Ley, Leiter der Deutschen Arbeitsfront (DAF), A. Hitler, hinter ihm Dr. Bodo Lafferentz, rechts Prof. Ferdinand Porsche.

▲ Reichsgebiet, Ostern 1936: Rast auf einem Waldweg in der Gegend von Stuttgart. Die ersten beiden, im Jahre 1935 gebauten VW-Versuchsmodelle, vorn ein Kabriolet (V 1) und dahinter die Limousine (V 2) während der Versuchsfahrt. Ganz links Ferry Porsche jun., rechts, mit der Flasche in der Hand, Versuchsingenieur Herbert Kaes.

Stuttgart-Zuffenhausen, 1938: Ein aus dem Fahrgestell des KdF-Wagens (VW-60) von der »SS-Fahrbereitschaft« selbstgebastelter »Kübelwagen«.

Stuttgart-Zuffenhausen, 1939: Prof. Porsche erklärt dem Gauleiter von Württemberg, Wilhelm Murr, die Konstruktion ◀ des KdF-Wagens.

2

Einige Tage vor dem Start zu den Versuchsfahrten der V 3-Wagen notiert Dr. Ferry Porsche, Sohn des Professors:
»Die Erprobung eines neuen Fahrzeugtyps erfolgt während der ersten Strecken (vielleicht bis 10 000 km) zweckmäßigerweise von einem festen Standort, und zwar von der Stelle aus, die sich mit dem Bau hauptsächlich beschäftigt hat. Es wird deshalb als Standort für die Erprobungsfahrten der ersten 3 Fahrzeuge Stuttgart vorgeschlagen. Wenn die Fahrzeuge nach jeder Tagesfahrt von etwa 600–700 km zum Standort zurückkehren, ist die dringend gebotene Möglichkeit vorhanden, irgendwelche auftretende Störungen nicht nur unverzüglich zu beheben, sondern auch durch entsprechende Maßnahmen sofort generell zu beseitigen.
Als Fahrtstrecke zur Erprobung der Fahrzeuge werden folgende drei Strecken vorgeschlagen:

Strecke I: 626 km	Stuttgart, Böblingen, Calmbach, Dobel, Herrenalb, Gernsbach, Baden-Baden, Oos, Rastadt, RAB-Bruchsal-Frankfurt-Bruchsal, Rastatt, Appenweier, Freudenstadt, Herrenberg, Stuttgart.
Strecke II: 554 km	Stuttgart, Bretten, RAB-Bruchsal-Frankfurt-Bruchsal, Oos, Appenweier, Kniebis, Freudenstadt, Stuttgart.
Strecke III: 487 km	Stuttgart, Bretten, RAB-Bruchsal-Frankfurt-Bruchsal, Gernsbach, Herrenalb, Dobel, Calmbach, Calw, Böblingen, Stuttgart.

Zunächst ist beabsichtigt, die Fahrzeuge täglich die Strecke I (626 km) mit dem Start gegen 6 Uhr morgens befahren zu lassen. In dieser Strecke sind Reichsautobahnen, Reichsstraßen und Landstraßen erster Ordnung abwechselnd enthalten. Die Steigungen bei Herrenalb, der sogenannte Viertälerweg, sowie am Kniebis dürften als besonders schwierig und steil bekannt sein. Die Strecke bei Herrenalb wird z. B. von Daimler-Benz zur Erprobung ihrer Fahrzeuge befahren. Es ist vorgesehen, den Beginn der Versuchsfahrten auf Montag den 5. Oktober 1936 festzusetzen. Bei einem Durchschnitt von zunächst 55 km pro Stunde würden für die Strecke I mit 626 km etwa 11 Stunden Fahrzeit im Tag benötigt. Zweckmäßigerweise lösen sich zwei Fahrer ab, so daß jedes Fahrzeug mit zwei Fahrern ständig besetzt ist. Nach Ablauf von etwa 10 000 km ist die Möglichkeit gegeben, die tägliche Leistung der Fahrzeuge durch Einschaltung der Strecken II und III beliebig zu vergrößern. Es können somit tägliche Fahrtleistungen von über 1000 km erreicht werden. Bei einer Gesamtstrecke von ca. 30 000 km je Wagen kann somit eine Ver-

suchsdauer von etwa 7–8 Wochen veranschlagt werden.
Zur einwandfreien Kontrolle der Fahrleistung werden in regelmäßigen Abständen eingehende Prüfungen durch das Forschungs-Institut, Professor Kamm, Stuttgart, angeregt. Diese Kontrollen sollen ähnlich wie bei der 30-Tagefahrt auf dem Nürburgring zu Beginn der Prüfungsfahrten, sowie nach je 6000 km erfolgen. Die Prüfungen erstrecken sich auf folgende Einzelheiten:
1. Beschleunigung,
2. Höchstgeschwindigkeit,
3. Bremsen.
Die Beobachtung der Fahrzeuge während der Prüfungsfahrten selbst erstreckt sich auf folgende Punkte:
1. Km-Leistung,
2. Durchschnittsgeschwindigkeit für die einzelnen Straßengruppen,
3. Maximalgeschwindigkeit (Reichsautobahn, Landstraße, Stadtstraße),
4. Kraftstoff- und Ölverbrauch,
5. Motortemperatur,
6. Abnützungserscheinungen usw.

Punkt 2 und 3 können dem Kienzle-Tachographen, welcher in jedem Fahrzeug eingebaut ist, entnommen werden.
Nach jeder Fahrt werden über den Verlauf der Fahrt, das Verhalten der Fahrzeuge, Störungen, Reparaturen, etwaige Veränderungen usw. Protokolle verfaßt, welche der Vertreter des RDA sowie die Dr. Porsche GmbH unterzeichnen. Diese Versuchsberichte bilden die Grundlage für den Gesamtbericht, der nach Abschluß der Fahrten ausgearbeitet wird. Zwischenberichte werden nicht herausgegeben, um das Endergebnis nicht irgendwie zu beeinflussen.
Etwaige auftretende Störungen während der Versuchsfahrt werden bei Fahrtunterbrechung sogleich behoben, damit die Ursache der Störung sofort erkannt werden kann und gegebenenfalls bei den anderen Fahrzeugen durch zweckentsprechende Abänderung vermieden werden kann. Auf keinen Fall ist es zulässig, die Prüfungsfahrten ohne Behebung fortzusetzen, da später die tieferen Ursachen nur erschwert oder gar nicht mehr feststellbar sind.
Die Belastung der Fahrzeuge während der Fahrten erfolgt wechselnd, wobei fehlende Personen durch Sandsäcke mit je 75 kg ergänzt werden. Die Mindestbelastung ist jeweils zwei Personen, die auf vier Personen einschließlich 30 kg Gepäck als Maximum erhöht werden kann. Die Aufteilung kann derart erfolgen, daß ein Drittel der Strecke mit einer Belastung von zwei Personen, ein Drittel mit drei Personen und ein Drittel mit vier Personen einschließlich 30 kg Gepäck in beliebiger Reihenfolge abwechselnd gefahren wird.
Die Organisation der Fahrt, die Aufnahme der Betriebsstoffe sowie etwaige Reparaturen erfolgen im Standort in einem besonders hierfür eingerichteten Raum, der unter genauer Kontrolle des RDA sowie der Dr. Porsche GmbH steht. An Betriebsstoffen können verschiedene Marken gefahren werden.
Nach Ablauf von je 3000 km wird jedes Fahrzeug einer möglichst weitgehenden Kontrolle unterzogen, um Mängel, welche während der Fahrt nicht aufgetreten sind, bereits im Entstehen ausfindig zu machen. Während der normalen Versuchsfahrten werden zur Erweiterung des Versuchsprogrammes größere Fahrten durch Deutschland vorgeschlagen. Die Besetzung der Fahrzeuge erfolgt dann jeweils von einem Fahrer, einem Vertreter des RDA und einem solchen der Dr. Porsche GmbH, die sich bei der Lenkung des Fahrzeuges beliebig ab-

wechseln. Die Fahrten können sich auf weitere Reisen nach Nord-, Mittel- und Süddeutschland oder in die Alpen erstrecken.«
Bei einem Tagespensum von 700 bis 1200 Kilometern werden die Wagen von sich ständig ablösenden Fahrerteams einer mehr als 50 000 Kilometer langen Zerreißprobe unterzogen, bis die drei Fahrzeuge zusammen etwa 160 000 Kilometer geschafft haben. Die Versuchsfahrten werden am 22. Dezember 1936 abgeschlossen.
Im Januar 1937 folgt ein Gutachten des RDA, das jede erdenkliche Schwäche des Wagens moniert. Darin heißt es: »Die Wagen hätten sich ›im allgemeinen‹ bewährt«, ein abschließendes Urteil könne man aber erst aufgrund eines weiteren Testes mit den nächsten dreißig vom RDA bestellten Probewagen fällen. Ihre Herstellung, erklärt der RDA, werde unter günstigen Bedingungen erfolgen, nämlich in den der Arbeitsgemeinschaft angeschlossenen Fabriken.«
Jetzt, nachdem die Herren des RDA fast zwei Jahre untätig zugesehen haben, wollen sie nun mit dreißig in eigenen Werken schnell produzierten Wagen bei Hitler Eindruck erwecken.
Das von der NS-Parteispitze eigens für die Vorbereitung der VW-Produktion gegründete Unternehmen heißt ›Gesellschaft zur Verbreitung des Volkswagens‹ (GEZUVOR). Geschäftsführer sind Prof. Ferdinand Porsche, Hitler-Intimus Jakob Werlin und Dr. Bodo Lafferentz, Mitarbeiter des Arbeitsfrontführers Robert Ley, dessen Organisation für die erforderlichen Mittel sorgen soll. Die Organisation Ley kann über ihre eigenen Finanzen hinaus auch über die Mittel der NS-Partei, ja sogar die der öffentlichen Hand verfügen.
Inzwischen entstehen fertigungsgemäß teils in Zuffenhausen, teils noch bei Mercedes in Untertürkheim zugerichtet, dreißig und dann noch einmal dreißig Prototypen, die auf Autobahnen und über Alpenpässe, auf einer Mammut-Strecke von insgesamt 2½ Millionen Kilometer gejagt werden.
Es bedarf nur eines Befehls aus Berlin, und schon kann die Porsche-Gezuvor-Gruppe auf das umfangreiche Motorenlager, die Garagen und die Wartungsmöglichkeiten der nahe Stuttgart gelegenen Kornwestheimer Wehrmachtskaserne zurückgreifen, wo man auch das Versuchs-Hauptquartier eingerichtet hat. Als Personal stellt Werlin SS-Männer bereit – bewährte, aus allen Teilen Deutschlands abkommandierte Fahrer – die unter der Leitung von Dr. Ferry Porsche Versuchsfahrten unternehmen. Die sogenannte Fahrbereitschaft ›VW‹ besteht anfänglich aus einem 120 Mann starken SS-Sturm, entsprechend einer Wehrmacht-Kompanie, geführt von SS-Hauptsturmführer Albert Liese.
Daß die ganze Angelegenheit keineswegs an die große Propagandaglocke gehängt wird, beweist schon die Erklärung, die jeder Teilnehmer der Versuchsfahrt ›W 30‹ – so der Tarnname der VW-Erprobungen – unterzeichnen muß. Den Schwätzern droht man mit der Gestapo und Strafverfolgung wegen Landesverrat:

Der Deutsche Volkswagen – Versuchsfahrt »W 30«

Erklärung

Mit meiner Unterschrift erkenne ich folgende Verpflichtungen für meine Mitarbeit bei der Versuchsfahrt W 30 als unbedingt bindend an:
1. Ich verpflichte mich, über alle die Versuchsfahrt angehenden Fragen und mir im Zu-

sammenhang damit zur Kenntnis kommenden Unterlagen, Konstruktions-Einzelheiten, Zeichnungen, Protokolle, Ergebnisse, Vorfälle und was es auch immer sei, strengste Verschwiegenheit gegen jeden, der außerhalb des mir von der Fahrtleitung der Versuchsfahrt W 30 bekanntgegebenen Bearbeiterkreises steht, zu beachten und mir gegebene Anweisungen zur Geheimhaltung getreulichst zu erfüllen.

2. Die Pflicht zur Verschwiegenheit besteht auch nach Abschluß meiner Tätigkeit bei der Versuchsfahrt W 30 weiter fort.
3. Ich verpflichte mich, alle mir bekannt werdenden Nachrichten, auch geringfügiger Art, die die Versuchsfahrt betreffen, oder deren Durchführung in irgendeiner Weise beeinträchtigen können, unverzüglich der Fahrtleitung W 30 schriftlich mitzuteilen.
4. Ich verpflichte mich, nachstehende Verbote strengstens zu beachten.
   Es ist verboten ohne ausdrückliche Genehmigung der Fahrtleitung W 30:
   a) die Mitnahme von irgendwelchen Unterlagen über die Versuchsfahrt W 30;
   b) die Anfertigung von irgendwelchen Vervielfältigungen, z. B. Photografien usw., sofern sie nicht von der Fahrtleitung W 30 ausdrücklich verlangt werden;
   c) irgendeiner Person außer der Fahrtleitung W 30 Einblick in die Fahrberichte oder in sonstige Unterlagen über die Versuchsfahrt W 30 zu gewähren;
   d) irgendwelchen Personen die Mitfahrt zu gestatten oder zu ermöglichen. Ausgenommen hiervon ist höhere Gewalt, dringende Gefahr oder notwendige Hilfeleistung, z. B. bei Unglücksfällen.
5. Ich verpflichte mich, alle etwa entsprechend den Dienstvorschriften während der Versuchsfahrt W 30 von mir angefertigten Lichtbilder als unentwickelte Negative der Fahrtleitung W 30 zu übergeben.
6. Ich habe davon Kenntnis genommen, daß ich für die Dauer meiner Mitarbeit bei der Versuchsfahrt W 30 gegen Unfälle während des Dienstes versichert bin.
7. Ich erkenne ausdrücklich an, daß ich keinerlei Regreßansprüche an den Veranstalter, an die Fahrtleitung der Versuchsfahrt W 30 oder deren Rechtsnachfolger stellen kann. Gerichtsstand für alle etwa anstehenden Streitigkeiten ist Stuttgart.
8. Ich verpflichte mich, vor und während des Dienstes keinerlei Alkohol in irgendwelcher Form zu mir zu nehmen.
9. Ich verpflichte mich, das Rauchen im Bereich der Versuchsbasis W 30 zu unterlassen.
10. Ich verpflichte mich, irgendwelche Tatsachen, Krankheiten usw., die meine Arbeitskraft oder mein Beobachtungsvermögen beeinträchtigen, der Fahrtleitung W 30 oder dessen Beauftragten vor Antritt des Dienstes mitzuteilen.
11. Ich verpflichte mich, allen Anordnungen der Fahrtleitung W 30 unverzüglich und sorgfältig Folge zu leisten und durch diszipliniertes Verhalten die reibungslose Durchführung der Versuchsfahrt W 30 zu sichern.
12. Ich verpflichte mich, die mir zur Kenntnis gebrachten Dienstvorschriften genauestens zu befolgen.
13. Ich habe zur Kenntnis genommen, daß die Fahrtleitung W 30 berechtigt ist, mich bei irgendwelchen Verstößen gegen die Dienstvorschriften oder durch mich hervorgerufenen Anlässen, welche die reibungslose Durchführung der Versuchsfahrt W 30

gefährden könnten, mit sofortiger Wirkung, ohne irgendwelche Regreßansprüche meinerseits, von der weiteren Mitarbeit auszuschließen.

14. Ich habe davon Kenntnis genommen, daß die Fahrtleitung W 30 verpflichtet ist, etwaige Verstöße gegen die auferlegte Schweigepflicht der Geheimen Staatspolizei mitzuteilen.

15. Es ist mir zur Kenntnis gebracht worden, daß ich bei Verstößen gegen die in dieser Erklärung aufgeführten Anordnungen Gefahr laufe, mich im Sinne des Gesetzes gegen unlauteren Wettbewerb in der Fassung vom 9. 3. 1932 (Reichsgesetzblatt 1932, Teil I, S. 122) und gegebenenfalls der §§ 88 ff. R. St. G. B., Abschnitt Landesverrat in der Fassung des Gesetzes vom 24. 4. 1934 (Reichsgesetzblatt 1934, Teil I) strafbar zu machen.
(Die vorstehend erwähnten Gesetze sind der Erklärung beigefügt; zum Teil im Auszug.) Vorstehende Erklärung kann jederzeit seitens der Fahrtleitung W 30 erweitert werden.

Stuttgart............. 1937 Unterschrift:

Das Unternehmen »W 30« wird der teuerste Test der Automobilgeschichte. Als die Versuche endlich beendet sind, stellt man fest, das deutsche Volk hat für seinen Volkswagen recht tief in die Tasche greifen müssen: Über 30 Millionen Reichsmark blieben auf der Strecke.
Nur für die Beseitigung der dabei festgestellten Mängel werden Mittel ausgegeben, für die private Automobilfabriken normalerweise ganze Modelle entwickeln. Und die Erprobungskosten des Volkswagens sind fast zehnmal so hoch wie die Kosten, die Privatunternehmen für ein neues Automobil aufzuwenden gewohnt sind.
Geld spielt bei diesem Geheimauftrag keine Rolle. Eine unendliche Reihe von Versuchen beginnt. Ehe Porsche auf den Vierzylinder-Boxermotor kommt, der später den VW antreiben soll, probiert er es mit einem Zweitakt-Zweizylinder, baut eine Dreizylinder-Reihe, experimentiert mit Doppelkolben, erprobt das Schnürle-System mit Taumelscheibenpumpe, konstruiert einen Zweizylinder-Viertakter und sogar einen Dieselmotor. Herauszufinden, in welcher Lage der Motor am besten eingebaut werde, kostet die Kleinigkeit von fünf Millionen Reichsmark. Auch für die Karosserie entwirft Porsche die verschiedensten Modelle; Holzkonstruktionen, Kunststoffgehäuse und Stahlkarossen. Allein die für das Produktions- und Versuchsprogramm dieser Wagen anfallenden Kosten erreichen astronomische Ziffern: Gegen Ende des Zweiten Weltkrieges steckt in dem VW-Werk ein Anlagekapital von 250 Millionen Mark.
Immerhin kostet 1 kg Volkswagen nur 1,60 Reichsmark – so preist Porsche seine Schöpfung an – im Vergleich zu Butter, für die man zu dieser Zeit pro kg 3,20 zahlen muß.
Im Frühsommer des Jahres 1937 erhält Hitler endlich seinen, dem deutschen Volk seit langem versprochenen Volkswagen, wenn auch erst in Gestalt eines Versuchsmodells. Was er bereits anläßlich der Automobilausstellung 1935 verkündet hat, geht nun in Erfüllung. Mit einem Heckmotor und dem Drehstab-Federungssystem, leichtem Gewicht und ungewöhnlichem Anzugsvermögen erweist sich der Volkswagen als ein äußerst zuverlässiges und zugleich erstaunlich billiges Hochleistungs-Gebrauchsfahrzeug.

Die Produktion kann beginnen, und hier ergeben sich neue Schwierigkeiten. Noch unter dem Eindruck seiner letzten Reise nach den USA will Porsche unbedingt auch den Volkswagen nach amerikanischer Manier fertigen. Es fehlen jedoch die deutschen Produktionsingenieure, die mit den entsprechenden *Know-How* vertraut sind. Es gibt nur einen Weg, das Problem zu lösen: diese Fachleute aus Amerika zu holen. Porsche reist abermals nach Detroit, wo auch ein Treffen mit Henry Ford vorgesehen ist, dem Hitler 1938 das Großkreuz des deutschen Adlerordens verlieh, weil seine Pionierarbeit »den Massen das Automobil geschenkt hatte«.
Man muß nur noch für das Vorhaben ein geeignetes Gelände finden, eine Fabrik bauen, sie vollständig einrichten und nicht zuletzt die Finanzierungsfrage lösen. Obgleich die Deutsche Arbeitsfront bereits über gewisse Eigenmittel verfügt, kann man sie natürlich nicht ausschließlich für dieses Fahrzeug verwenden.
Das Problem wird mit Hilfe der »Kraft-durch-Freude« (KdF)-Organisation auf verblüffende Weise gelöst. KdF kassiert jährlich zwischen 54 und 80 Million Reichsmark, deren Verbuchung geheim bleibt. Die Verwendung der Mittel ist dem Belieben der Partei überlassen: Mit einem Wort, es handelt sich um einen »politischen Geheimfond«. KdF übernimmt die Schirmherrschaft für das Volkswagenprojekt. Man klügelt einen raffinierten Sparvertrag mit Wochenraten von 5 Mark aus, den Dr. Ley in seiner Kölner Rede vom 1. August 1938 der deutschen Arbeiterschaft bekanntgibt. Für diesen gigantischen Sparvertragsplan legen schließlich 336 668 Deutsche insgesamt 280 Millionen Reichsmark auf die hohe Kante. Die Organisation »Kraft-durch-Freude« deponiert die Beiträge der Werktätigen bei der Bank der Deutschen Arbeit, und als Berlin fällt, liegt diese Summe unberührt auf dem Konto, die dann allerdings von den eindringenden Russen beschlagnahmt wird.
Am 26. Mai 1938, es ist Himmelfahrtstag, legt Adolf Hitler in der Nähe von Fallersleben in Niedersachsen den Grundstein zum Volkswagenwerk. Er selbst hält die Rede und bestimmt, daß der in dem neuen Werk zu produzierende Wagen den Namen »KdF-Wagen« trage. Zum 15. Oktober 1939 sollen die ersten 500 Autos vom Band laufen und noch vor Jahresende die Serienproduktion in vollem Gange sein.
Das Werk, so prophezeit Hitler, werde die Fordfabriken in Michigan und ihre Produktion in den Schatten stellen. Und tatsächlich ist das Nachkriegs-Volkswagenwerk in seiner heutigen Gestalt die größte Automobilfabrik der Welt.
Die Fachzeitschrift »Motor-Kritik« notiert begeistert im Jahre 1938: »Langsam beginnen sich die Schleier um den konstruktiven Aufbau des KdF-Wagens zu lüften. Bislang konnte man wohl den äußeren Aufbau in Augenschein nehmen, bereitwilligst wurde der Deckel zum Kofferraum geöffnet, man konnte auch bei Veranstaltungen den Lauf der Maschine abhören oder sich von dem Temperament des Wagens ein Bild machen, wenn man auf der Autobahn einem Versuchswagen begegnete.
Die Haube, die den Motorraum überdeckt, war in allen Fällen hermetisch verschlossen, und neugierige Fragen prallten wirkungslos an der stoischen Ruhe der Begleitmannschaften ab.
Als Weihnachtsgeschenk für die vielen Wißbegierigen erscheinen jetzt die ersten Bilder und näheren technischen Angaben über den KdF-Wagen. Der Boxermotor liegt hinter der Hinterachse, das Vierganggetriebe vor der Hinterachse.

Das Kurbelgehäuse des Vierzylindermotors dient gleichzeitig als Fundament für das verschiedene Zubehör (Kühlluft-Gebläse, Lichtmaschine usw.). Der Mittelträger als Rückgrat des ganzen Fahrzeuges ist an seinem hinteren Ende gegabelt, um das Antriebsaggregat aufzunehmen. Eine sickenartige Profilierung des Fahrzeugbodens verleiht diesem in Verbindung mit einer Rahmeneinfassung eine große Steifigkeit trotz größter Gewichtseinsparung.«

Jedoch ist kein einziger Volkswagen eines privaten Besitzers je über die Straßen des Dritten Reiches gerollt. Bis zur Umstellung auf Rüstungslieferungen hat man lediglich 210 KdF-Wagen gebaut, die, nicht einmal in Wolfsburg selbst hergestellt, ausschließlich an NS-Dienststellen oder Parteifunktionäre geliefert werden; die KdF-Sparer gehen dabei völlig leer aus.

Reichsgebiet, Winter 1940/41: Der Kraftfahrzeug-Park einer Luftwaffen-Einheit mit nagelneuen VW-82 Kübelwagen.

Reichsgebiet, Winter 1940/41: Der Kraftfahrzeug-Park einer Luftwaffen-Einheit. Ein Wachtposten zwischen den abgestellten VW-82 Kübelwagen.

Reichsgebiet, Winter 1940/41: Die Fahrzeugkolonne mit VW-82 einer Luftwaffeneinheit verläßt in Marschordnung die Kaserne.

Reichsgebiet, Frühling 1941: Die Troßkolonne einer Kradfahrer-Einheit der Division »Großdeutschland« mit ihren VW-82 auf dem Marsch.

Nord-Afrika, Cyrenaika, Juni 1941: Ein VW-82 des DAK nach dem Sandsturm. Das Verdeck ist verschnürt, die Kühlerschlitze im hinteren Kastenteil abgedichtet, die Hinterräder notdürftig mit Kistenbrettern verschalt. Sowohl die Rückblickfensterscheibe im Verdecküberzug, wie auch das mit Steckzapfen und Bolzen am oberen Türrahmen befestigte Steckfenster sind durch die Einwirkung des feinen Sandes matt geschliffen.

Berlin, Sommer 1941: Ein Leutnant und seine Männer posieren bei einer fröhlichen Fahrt in einem VW-82 für ein Erinnerungsfoto.

Nord-Afrika, Sommer 1941: Der Fahrer eines Kübelwagens während der wohlverdienten Mittagspause.

Nord-Afrika, Libyen, Sommer 1941: Ein Kübelwagen der Luftwaffe vor einem italienischen Wüstenfort an der Mittelmeerküste in der Nähe von Misurata. Rechts neben dem Fort ein Marine-Signalmast.

Nord-Afrika, bei Tripolis, Sommer 1941: Auf der Fahrt mit einem VW-Kübelwagen. Vorn, längs des Windschutzscheiben-Rahmens der Winker. Ganz rechts unten der Griff des Suchscheinwerfers. Links hinter dem Beifahrer das hintere Steckfenster mit Steckzapfen. Die Besatzung des VW-82 trägt die khaki-braunen Tropenhemden, der Beifahrer hat den Tropenhelm, die wegen ihrer Nachteile recht unbeliebte Kopfbedeckung des DAK, auf.

Nord-Afrika, Libyen, Sommer 1941: An einer Wasserstelle in der Wüste. Ein italienischer Soldat bedient die Handpumpe des artesischen Brunnens. Der Fahrer eines VW-82 der Luftwaffe füllt die 20-Liter-Einheits-Kanister mit Wasser.

Nord-Afrika, Libyen, Sommer 1941: Vor der weiteren Fahrt noch einen Blick auf die Karte. Die Besatzung des VW-82 trägt die Feldmützen des DAK. Aus diesen Mützen entsteht später die Einheitsfeldmütze des deutschen Heeres, Modell 1943. Der Soldat auf dem hinteren Kastenteil trägt zur Tropenanzug-Feldbluse kurze Hosen und die für das DAK typischen verschnürten Segeltuchstiefel. Im Wagen sitzen zwei englische Kriegsgefangene. Die umgeklappte Windschutzscheibe befindet sich in einer Schutzhülle. Auf dem unteren Scheibenrahmen sind beide Scheibenwischermotoren zu sehen, links daneben der Suchscheinwerfer mit Abblendkappe. Zwischen Windschutzscheibe und Ersatzrad liegt der mit Lappen und Gummiringen notdürftig vor Sandeinwirkung abgeschirmte Einfüllstutzen des Kraftstoffbehälters.

Nord-Afrika, Tripolis, Sommer 1941: Radwechsel inmitten der Wüste. Man sieht die rechte Hinterradnabe und Bremstrommel des VW-82. Unter der Hinterachse der Wagenheber.

Griechenland, Sommer 1941: Auf dem Gelände einer ehemals griechischen Kaserne bei Megara, östlich von Athen. Eine Kfz Schützen-Einheit des gerade aufgestellten Deutschen Afrika-Korps ist zum Appell angetreten. Im Hintergrund mehrere VW-82, mit MG 34 auf Schwenkarm.

Megara-Kaserne: Aufsitzen nach dem Appell. Die Einheit trägt oliv-grüne Tropenhelme, leichte Drillich-Tropenanzüge und kurze Schnürstiefel.

Megara-Kaserne: Die Soldaten eilen zu ihren Fahrzeugen. Links, auf einem behelfsmäßigen Wagenanhänger, ein leichtes Geschütz.

Megara-Kaserne: Die Besatzungen nehmen in ihren VW-82 Kübelwagen Platz.

Megara-Kaserne: Vor dem Aufbruch zur Fahrübung in Marschordnung. In der Mitte ein Mercedes-Benz, Typ L 1500 A mit einem leichten Geschütz am Anhänger.

Megara-Kaserne: Der Gruppenführer gibt mit der Kelle das Signal zur Übungsfahrt.

Megara-Kaserne: Fahrübungen der Einheit auf dem Kasernengelände.

Megara-Kaserne: Eine Pause zwischen den Übungsfahrten auf dem Kasernengelände.

Megara-Kaserne: Übungen auf dem Kasernengelände. Vorn, der Einheitsführer, erteilt dem Gruppenführer Unterricht in der Handhabung der Kelle.

Megara-Kaserne: Vor dem Aufsitzen zu Geländefahrtübungen. Vorn der Gruppenführer.

Megara, Sommer 1941: Geländefahrtübungen auf einem Feldweg zwischen Megara und Elevsis.

3

Die Umwandlung des Porsche-Volkswagens in einen Kriegswagen wird am 17. Januar 1938 aktenkundig. An diesem Tage verfaßt der SS-Hauptsturmführer Albert Liese, Leiter der SS-Fahrbereitschaft VW, einen Bericht über seinen Besuch beim Chef des Heeres-Waffen-Amtes, General der Infanterie Liese, am Freitag, dem 14. Januar 1938.

»Die Besprechung« – notiert SS-Liese, »erstreckte sich auf eine Darlegung des Fahrgestells des VW anhand der mitgebrachten Zeichnung. Herr General Liese äußerte den Wunsch, man möge ihm den Volkswagen gelegentlich vorführen, damit er einen Eindruck von dem Fahrzeug erhalte. Oberstleutnant Fichtner, der schon bei den Vorführungen vor den Stabsoffizieren der verschiedenen Wehrmachtsteile in der Taubertstraße im Dezember vorigen Jahres zugegen war, erhielt vom Herrn General Liese den Auftrag, die seitens des Heeres-Waffen-Amtes gestellten Forderungen baldmöglichst bekannt zu geben, damit ein Fahrzeug entsprechend ausgerüstet werden könnte.

Es wurde mit Herrn General Liese vereinbart, daß die von uns weiter entwickelten Spezialaufbauten für Gelände – z. B. auf dem Truppenübungsplatz in Münsingen – erprobt werden sollen, um etwaige Kinderkrankheiten auszumerzen.

Anschließend an die Besprechung mit Herrn General Liese fand eine weitere Besprechung mit Herrn Oberstleutnant Fichtner statt. Diese hatte zum Ergebnis, daß uns im Laufe dieser Woche – voraussichtlich am Mittwoch oder Donnerstag – nach Überprüfung der Zeichnung die Wünsche des Heeres-Waffen-Amtes bezüglich des Aufbaues und der Geländegängigkeit des Fahrzeugs bekannt gegeben werden. Bei dieser Besprechung, die wieder in Berlin stattfinden soll, werden uns die erforderlichen Unterlagen an die Hand gegeben, mit denen wir zusammen mit der Karosseriefabrik ›Trutz‹ in Coburg – die Firma baut seit Jahren Aufbauten für die Wehrmacht für alle Zwecke – einen den Anforderungen des Heeres-Waffen-Amtes entsprechenden Aufbau entwickeln können.

Nach den Ausführungen des Herrn Oberstleutnant Fichtner kommt für den Volkswagen eine militärische Verwendbarkeit insofern in Frage, als das Fahrzeug gemäß seiner Stärke mit 3 Mann und einem schweren Maschinengewehr besetzt werden kann. In dem Begleitschreiben an das Heeres-Waffen-Amt wurde das Gesamtgewicht des Fahrzeugs mit 950 kg angegeben. Wenn auch die Belastung der Fahrzeuge bei der 50 000-km-Fahrt im Jahre 1936 bis zu 1100 kg auf der Straße betragen hat, so ist es nicht ratsam über eine Belastung von 950 kg hinauszugehen, da die Fahrzeuge be-

kanntlich bei Geländefahrten wesentlich höher beansprucht werden und daher nicht voll belastet werden können.
Für drei Mann Besatzung und ein schweres Maschinengewehr rechnet das Heeres-Waffen-Amt einschließlich Munition insgesamt 400 kg. Bei einem Chassis-Gewicht von 390–400 kg würden also 150–160 kg für den Aufbau zur Verfügung stehen. Nach den Erfahrungen des Heeres-Waffen-Amtes wird mit einem Aufbaugewicht in dem vorliegenden Fall von etwa 200 kg gerechnet. Es wird uns anheim gestellt, dieses durch die weitgehendste Verwendung von Leichtmetall möglichst herabzusetzen.
Zum Schluß der Besprechung fragte ich Herrn Oberstleutnant Fichtner, der für die Beurteilung des Fahrzeugs über die militärische Verwendbarkeit in gewissem Umfange zuständig ist, was er nach der vorliegenden Zeichnung von der militärischen Verwendbarkeit des Volkswagens halte.
Oberstleutnant Fichtner bemerkte daraufhin, daß etwa nach der vorliegenden Zeichnung das Fahrzeug wesentlich sympathischer aussehe. Bei der Besprechung wurde weiter eine Zeichnung eines normalen geländegängigen kleinen PKW vorgelegt, die zeigte, daß der Volkswagen mit Radstand und Spur durchaus in der Mitte der vom Heeresamt gestellten Forderungen liegt. Bezüglich der Bodenfreiheit besteht seitens des Heeres-Waffen-Amtes eine Forderung von 240 mm unter der Mitte eines Fahrzeuges bei dem bekannten 8 m Krümmungsradius. Als ich Herrn Oberstleutnant Fichtner andeutete, daß man diese Bodenfreiheit auch beim Volkswagen ohne Schwierigkeiten erreichen könne, bemerkte er, daß es auf die paar Millimeter nicht so sehr ankomme. Die Hauptsache sei, daß man das Bestmöglichste aus dem vorhandenen Fahrzeug für die militärische Verwendbarkeit heraushole. Bei derselben Gelegenheit bat ich auch, uns die Konstruktionsbedingungen für den Bau von kleinen PKW auszuhändigen. Dieses gelang aber nicht, da eine solche Vorschrift scheinbar im Heeres-Waffen-Amt nicht vorzufinden war.
Der Gang der weiteren Entwicklung ist folgender:

1. am 19. oder 20. Januar 1938 Besprechung im Heeres-Waffen-Amt in Berlin.
   Bekanntgabe der Bedingungen des Heeres-Waffen-Amtes.
2. Entwicklung des Geländereifens für den Volkswagen.
3. Entwicklung eines militärischen Aufbaues zusammen mit der Firma Trutz, Coburg.
4. Vornahme von Versuchsfahrten im Gelände seitens der Dr. Ing. h. c. F. Porsche KG.
5. Versuche des Heeres-Waffen-Amtes, wobei die Dr. Ing. h. c. F. Porsche KG entsprechend beteiligt ist.

gez. SS-Hauptsturmführer Albert Liese.«

Schon einige Tage später schreibt SS-Hauptsturmführer Liese einen neuen Aktenvermerk, diesmal über eine Besprechung im Heeres-Waffen-Amt, die am Mittwoch, dem 26. Januar 1938 stattfindet:
»Die Wehrmacht beabsichtigt, das Chassis des Volkswagens für den Transport von 4 Soldaten in kriegsmäßiger Ausrüstung beim Gewicht von je 100 kg zu verwenden. Die Gewichtsverteilung setzt sich somit folgendermaßen zusammen:

Zulässiges Gesamtgewicht im Gelände	950 kg
Chassis-Gewicht mit Geländebereifung	400 kg
Zulast für Soldaten mit kriegsmäßigem Gepäck	400 kg

Für den Aufbau verbleibt damit ein Gewicht von 150 kg.
Die bisher bei der Wehrmacht in Betrieb befindlichen Kübelsitzaufbauten auf leichtem PKW betragen durchweg 300–400 kg. Wenn in dem vorliegenden Fall ein Gewicht des Kriegsaufbaues von 150 kg erreicht werden soll, müssen vollständig neue Wege gegangen werden. Aus diesem Grunde läßt das Heeres-Waffen-Amt bei der Entwicklung des für den VW bestimmten Kübelsitzaufbaues den Konstrukteuren vollständig freie Hand. Wir erhalten in den nächsten Tagen ausführliche Zeichnungen, welche über die seitens des Heeres-Waffen-Amtes geforderten Abmessungen der Sitze, des Radraumes, der Windschutzscheibe, des Verdecks usw. Aufschluß geben.
Bei der Besprechung im Heeres-Waffen-Amt ergibt sich, daß an der Entwicklung des Volkswagens mit 2 Mann und einem Maschinengewehr nach amerikanischem Vorbild die Truppe wegen des niedrigen Aufbaues besonders interessiert ist, da es ein solches Fahrzeug mit Heck-Motor und niedriger Chassis-Höhe bisher noch nicht gibt.
Die Techniker des Heeres-Waffen-Amtes glauben aber, dem amerikanischen Vorschlag keine große Bedeutung beimessen zu können. Aus diesem Grunde nimmt Liese nach vorheriger Rücksprache mit Dr. Porsche sofort Verhandlungen mit dem Chef der SS-Leibstandarte Adolf-Hitler, SS-Obergruppenführer Dietrich, auf, der die Entwicklung eines solchen Fahrzeuges ebenso begrüßt, wie die Truppe. Es ist die Möglichkeit geboten, mit den zuständigen Stellen der Leibstandarte Adolf Hitler einen solchen Aufbau versuchsweise durchzuführen und auf seine Verwendung hin zu prüfen.
SS-Obergruppenführer Dietrich sieht die Bedeutung dieses Wagens besonders in folgenden Punkten:
1. Geringe Bauhöhe,
2. leichtes Gewicht,
3. große Stückzahl,
4. geringe Umbaukosten vom normalen Fahrzeug zum Kriegsfahrzeug,
5. verschwindend geringe Gesamtkosten gegenüber dem bisher von der Wehrmacht entwickelten Fahrzeug.

Der Vorschlag des Herrn Dr. Porsche für die Geländebereifung des Volkswagens dieselbe Felgengröße zu verwenden, wie beim normalen Reifen, wird vom Heeres-Waffen-Amt anerkannt. Maßgeblich für die Gründe sind die große Stückzahl des VW sowie die Materialersparnis. Aus beiden Gründen ist es unter keinen Umständen zulässig, eine andere Felgengröße für Wehrmachtszwecke zu verwenden.

gez. Liese, SS-Hauptsturmführer.«

Die Sache mit dem Militär-VW, der unter dem Namen VW-Kübelwagen in die Geschichte des Zweiten Weltkrieges eingeht, rollt immer schneller. Schon am Dienstag, dem 1. Februar 1938, wird der Volkswagen in Berlin-Grunewald, Tauberstr. 1, dem Chef des Heeres-Waffen-Amtes, General der Infanterie Liese, vorgeführt. Anwesend sind dabei:
Dipl.-Ing. Ministerialdirigent Poller und
der Adjutant des Generals, Major Herfurth,
vom Prüfwesen 6 der Chef der Abteilung
Oberstleutnant Fichtner,
vom Heeres-Waffen-Beschaffungsamt
Oberst Stud und
Oberst Zuckertort.
Am Nachmittag des 1. Februar findet eine Besichtigung im Heeres-Waffen-Amt durch

den General der Artillerie, Professor Dr. phil. h. c. Dr.-Ing. Becker, dem Leiter des Prüfwesens beim Heeres-Waffen-Amt statt. Ferner sind zugegen: Adjutant des Generals, Major Dipl.-Ing. John und Major von Kolb.

Am Mittwoch, dem 2. Februar 1938, startet eine Probefahrt mit Herrn General Becker und seinem Adjutanten Major John zum Truppenschießplatz nach Kummersdorf.

Anschließend an diese Vorführung vor General Liese werden die Fahrzeuge dem General der Panzertruppen Lutz in Berlin W 35, Kurfürstenstr. 78, vorgeführt.

Der Zeit der fieberhaften Versuche folgen Arbeit, unzählige Probefahrten und Besprechungen. Anfang November 1938 ist es endlich so weit: »Am Donnerstag, dem 3. 11. 1938« – notiert SS-Hauptsturmführer Liese – »wird das erste Versuchsmodell des Kübelsitz-VW im Heeres-Waffen-Amt vorgeführt.

Zugegen sind:

General Becker, der Referent für Aufbauten im HWA und andere Herren. Mit dem ersten Modell waren die Herren des Heeres-Waffen-Amtes einverstanden. Änderungen bzw. Wünsche wurden auf Anfrage nicht vorgetragen.«

Es wird vereinbart, daß Versuchsfahrten im Gelände baldmöglichst erfolgen sollen.

Zur Durchführung von Vergleichsfahrten im Gelände wird am Montag, dem 14. November, in Münsingen ein Einheits-PKW (Vierrad-Antrieb, Vierrad-Lenkung) zur Verfügung stehen. Beide Fahrzeuge werden gemeinsame Versuchsfahrten im Gelände durchführen.

Damals hieß dieser Wagen-Typ im amtlichen Sprachgebrauch noch Kübelsitz-VW. Die Sitze im kastenartigen Mannschaftswagen – dem heutigen Schalensitz ähnlich – nannte man anfangs Kübelsitze, und erst das Landser-Latein machte aus dem Kübelsitzwagen einfach den Kübelwagen. Porsche, dem es nicht gegeben war, dem Volk einen Wagen zu bauen, sorgte wenigstens dafür, daß die Feldgrauen mit seinem VW-Kübelwagen einen vorzüglichen, leichten Gelände-PKW bekamen. Zwar hat der Kriegswagen keinen Allrad-Antrieb, doch seine Bodenfreiheit, die einfache, aber robuste durchgehende Karosserie samt Heckmotor lassen diesen Nachteil leicht vergessen. Und Porsche ist der erste, der auf die Idee kommt, daß ein Geländewagen keineswegs ein schweres und kompliziertes Vehikel sein muß. Im Gegenteil, ein einfacher, leichter Wagen hat viel mehr Überlebenschancen. Auch der beinahe zierlich wirkende, luftgekühlte Ein-Liter-Motor erweist sich als völlig richtig.

»Im Krieg ist Luftkühlung gut, denn im Krieg wird geschossen, und es können Löcher im Wasserkühler entstehen. Und wenn beim Luftgekühlten etwas nicht stimmt, dann läuft der Motor ungekocht weiter und duldet still in sich hinein.« – meint der große Meister.

Die VW-Wehrmachtsausführung, den Kübelwagen, stellt man erstmalig dem Publikum auf der Wiener Messe 1939 vor. Der neue, für die Bedürfnisse der Wehrmacht entworfene VW-Geländewagen bekommt die Porsche-Typenbezeichnung ›62‹. Die Prototypen des VW-62, im Jahre 1939 gebaut, mit 3- bzw. 4sitzigen offenen Aufbauten wiegen 642 kg, jedoch ist das OKH trotz der größeren Bereifung (5,00–18) mit ihrer Geländegängigkeit nicht ganz zufrieden. Erst die von Porsche weiterentwickelte Ausführung VW-82, der in seiner endgültigen Form bekannte Kübelwagen, besitzt eine größere Bodenfreiheit. Durch den Einbau eines Zwischengetriebes an den Hinterrädern wird die Gesamtübersetzung entscheidend geändert, auch ein Sperrdifferential ist vorgesehen. Das Gesamtgewicht ohne Last: 750 kg.

Dieses seltsam anmutende Automobil, das, ab 1940 gebaut, seinen Namen »Kübelwagen« zu Recht trägt, ist ein offenes, kastenförmiges Fahrzeug mit VW-ähnlicher, abgeflachter Vorderpartie, auf die man das Ersatzrad montiert hat. Der Porsche-Kübel- und -Schützenwagen bewährt sich als das glänzendste Kriegsauto. Der luftgekühlte Porsche-Motor (mit einem Ölkühler versehen, den bis dahin nur hochwertige Sportfahrzeuge aufwiesen) kann in Rommels Wüste nie zu heiß, in der Halbarktis Nordnorwegens und in den russischen Winterschlachten nicht zu kalt werden. Der Frostschutzmittel-Nachschub wird dadurch überflüssig, und der Wagen ist selbst in den Gefilden Rußlands noch einsatzbereit, wo es kein Kühlwasser mehr gibt.

Daß sich die Wehrmacht bei der Wahl ihres Standardfahrzeuges nur zögernd für den Kübelwagen entscheidet, liegt nicht zuletzt an dem starken Druck, den die Automobilhersteller zur Wahrung ihrer Anteile am Rüstungsgeschäft auf sie ausüben. Keiner von ihnen will sich mit einem Monopol des staatseigenen Volkswagenwerkes für Militärpersonenfahrzeuge abfinden. So z. B. Carl Borgward, Inhaber der gleichnamigen Automobilfabrik, Wehrwirtschaftsführer für Norddeutschland und damit Herr über die gesamte Produktion von Wehrmachtsfahrzeugen in Niedersachsen, befürchtet durch eine Typenbeschränkung die Lahmlegung seines Betriebes.

Als der Krieg ausbricht liegen weder bei der Werksleitung in Wolfsburg noch in Berlin Pläne bereit, das Werk der Kriegsführung nutzbar zu machen. In dem Glauben an einen baldigen Sieg zögert man mit dessen Umstellung auf Rüstungslieferungen und wartet auf das Anlaufen der Volkswagenproduktion.

Nun, als endlich das VW-Werk 1940 zu produzieren beginnt, werden – so die Ironie des Schicksals – die rechteckigen Kübelwagen-Aufbauten in der Berliner Ambi-Budd Fabrik, einer Tochterfirma des gleichnamigen amerikanischen Konzerns, hergestellt. Nur das Kübelfahrgestell baut man in einer der Ecken des monströsen VW-Werkes, da in den anderen Sektoren Minen, Fertigteile für Ju 88-Bomber und allerlei Kriegszeug angefertigt werden, wie z. B. 1,5 Millionen Behelfs-Heizöfen, an denen sich die Landser im bitterkalten Rußland wärmen können.

Dann setzen alliierte Luftangriffe ein. Das Werk, ohne jeden Schutz oder Tarnung im flachen Land erbaut, bietet am Mittellandkanal, mit seinen wuchtigen, markant rechteckigen Werkhallen eine wahre Zielscheibe für Bombenflugzeuge.

Eine Gutachterkommission des Strategischen Bomberkommandos der USA, die nach der deutschen Kapitulation das Kriegspotential des Volkswagenwerks und seine erlittenen Bombenschäden untersucht, stellt dabei fest, daß »bei Kriegsbeginn weder die Werkleitung noch die zuständigen Behörden in Berlin Vorkehrungen getroffen haben, das Werk der Kriegsführung nutzbar zu machen. Die Reichsregierung rechnete allem Anschein nach mit einem kurzen Krieg und beschloß daher, das Werk und seine Einrichtungen für die Nachkriegsproduktion von Kraftwagen in seinem damaligen Zustand zu belassen. Das geht auch daraus hervor, daß in dem Werk – wie etwa 1941 – die Kapazität nur zu 20 bis 25 % ausgenutzt wurde. Den ganzen Krieg über hat man es zu keiner Zeit mit mehr als 50 % Produktion belegt«.

Genau besehen, stellte es erstaunlich wenig Wehrmachtsfahrzeuge her. Von 1940 bis 1945 kamen etwa 55 000 VW-Kübelwagen und

15 000 VW-Schwimmwagen aus Wolfsburg. Das Volkswagenwerk erreichte während keiner Phase des Krieges auch nur annähernd seine volle Produktionskapazität.

Griechenland, Sommer 1941: Die Kfz-Schützen-Einheit des DAK auf der Fahrt durch das Städtchen Elevsis in der Nähe von Athen.

Griechenland, Sommer 1941: Ein Erinnerungsfoto von der Geländefahrtübung. Im Tal das Städtchen Megara.

Reichsgebiet, Sommer 1941: Der nagelneue VW-82 eines Geschwader-Kommandeurs.

Nord-Afrika, Libyen, Herbst 1941: Einer der seltenen besinnlichen Augenblicke während der Einsatzpause. Die beiden Soldaten sind in ihre Brieflektüre vertieft. Der VW-82/trop ist mit Kronprinz-Sandreifen und Radzierkappen ausgestattet. Das Ersatzrad auf dem vorderen Kastenteil ist der Glätte nach eher einem Schlauch ähnlich. Der Buchstabe R auf dem rechten vorderen Kotflügel steht für die 2. Fallschirmjäger-Division.

Nord-Afrika, Libyen, Herbst 1941: Plauderei am Rande der Wüstenstraße. Der Innenanstrich der vorderen rechten Seitentür zeigt am besten den Unterschied zwischen der üblichen dunkelgrauen Tarnfarbe des Heeres und dem sandgelben Tarnanstrich des DAK.

Irgendwo in der Cyrenaika, September 1941: Eine kurze Rast für Mensch und Wagen inmitten der Wüste. Ein VW-Kübelwagen 82/trop.

Nord-Afrika, Spätherbst 1941: Ein VW-82 Kübelwagen. Links auf dem Türblech das Zeichen des DAK. Man sieht besonders gut die beiden Türgriffe und -scharniere, Karabinerhalter, Scheibenrahmen und das Verdeckgestell mit Überzug.

Nord-Afrika, Libyen, Herbst 1941: Geschäfte an der Wüstenstraße. An der linken Seite des vorderen Kastenteils neben der Windschutzscheibe die Halterung des Suchscheinwerfers.

Nord-Afrika, Libyen, Herbst 1941: Ein VW-82 neben einem Luftwaffenstützpunkt unweit des Städtchens El Nofila. Rechts ein erbeuteter Citroën 11 CV Normale, Baujahr 1938.

Nord-Afrika, Libyen, Herbst 1941: Verschnaufpause für die Soldaten und den VW-82 am Rande des Schlachtfeldes. Auf dem hinteren Sitz eine Kiste mit Spatenbräu Bier aus München.

Nord-Afrika, Libyen, Herbst 1941: Ein VW-82 der Luftwaffe. Über dem vorderen und hinteren Kotflügel die nachträglich angebrachten Halterungen für den 20-Liter-Einheitskanister und einen Spaten. Im Hintergrund eine für das DAK lebensnotwendige Einrichtung: Die Feld-Instandsetzungswerkstatt.

Nord-Afrika, Cyrenaika, Herbst 1941: Die Besatzung eines VW-Kübelwagens posiert für ein Erinnerungsfoto. Auf dem Ersatzrad das Fliegererkennungszeichen, am linken Kotflügel das nachträglich montierte Gestell für den 20-Liter-Einheitskanister.

Nord-Afrika, südlich von Tripolis, Herbst 1941: Ein Kübel-VW mitten auf dem Gebel-Nefusa-Plateau. Der Hirtenjunge bestaunt das Gewehr K 98 b.

Nord-Afrika, Libyen, Herbst 1941: Der VW-Kübelwagen Typ 82 vor einem Verwaltungsgebäude am Rande der Wüste.

Nord-Afrika, Libysche Wüste, Herbst 1941: Ein Kübelwagen der Luftwaffe mit Sandrädern fährt eine Düne hinunter.

Raum Kiew, Herbst 1941: Rast auf dem Vormarsch. In der Mitte ein VW-82, links ein schwerer Einheits-Pkw, rechts ein Kübelwagen Horch, Typ 830 R.

Ostfront, Herbst 1941: Selbst die Räder des Kübelwagens drehen auf der vom Regen aufgeweichten Rollbahn durch.

Ost-Feldzug, Südfront, Herbst 1941: Der hohe Besuch ist mit einem VW-82 im Schlamm steckengeblieben. Rumäniens Marschall Antunescu bei einer Inspektionsfahrt.

4

Die ganze Großdeutsche Presse von der Maas bis an die Memel, von der Etsch bis an den Belt bejubelt den VW-Kübelwagen. Selbst im kleinen Asch, im Sudetenland, erscheint am 9. 8. 1941 in der dortigen *»Ascher Zeitung«* der Aufsatz eines begeisterten Redakteurs:

»Der geländegängige Volkswagen im Einsatz

Der KdF-Wagen – auf Wunsch des Führers geschaffen, um jedem schaffenden Deutschen Erholung und Freude zu bringen – hat jetzt im Kriege bei der Wehrmacht als geländegängiger Volkswagen neue Aufgaben gefunden. An allen Fronten hat er seine viel gepriesenen Fahreigenschaften nicht nur bewiesen, sondern bei weitem übertroffen. Wochenschau, sowie Bild- und Wortberichte der Propagandakompanien zeigen ihn im Einsatz und allen Soldaten, die ihn kennen, ist er ein unentbehrlicher guter Kamerad geworden.

». . . Vorbei hasten wir an Kolonnen. Der Volkswagen brummelt und pruscht brav seinen Weg. Der Volkswagen, luftgekühlt, klein, zäh, mit dem wüstenfarbenen Anstrich, mit dem Motor hintendrein, bewährt sich hier einzigartig und frißt, was die Wüste ihm entgegenwirft; schnauft, spuckt einmal, aber er wird mit allem fertig, und wir klopfen ihn am Abend auf die Krippe, wie er es verdient hat und wie es jedermann in Deutschland wissen soll. Fahr mit ihm wohin Du willst, mein Junge. Wischst Du ihm den Sand aus Hals und Röhren, spürt er deine gute Hand – er läßt dich nicht im Stich.«

So schreiben unsere Kriegsberichter über ihre Erlebnisse mit dem Volkswagen in der Wüste. Ihre Berichte sind zugleich der beste Beweis für die wirkliche Volkstümlichkeit und Beliebtheit unseres Volkswagens.

Ob auf den Straßen der Heimat oder im tiefsten Sand der libyschen Wüste, überall hat der KdF-Wagen als geländegängiger Volkswagen seine Bewährungsprobe vorzüglich bestanden. Luftgekühlter Motor, leicht und doch stabil, das sind die wichtigsten Eigenschaften, die ihn auch in der Hitze Afrikas zum besten Kameraden der Soldaten werden ließen, denn die schwierigen klimatischen und Geländeverhältnisse in Afrika erfordern nicht nur den ganzen Mann, sondern vor allem auch erstklassige Kraftfahrzeuge, um die gewaltigen Strekken in der unwegsamen und unendlichen Wüste schnell und sicher bezwingen zu können. Natürlich mußte der KdF-Wagen im Kriegseinsatz auch sein schönes Kleid ablegen und sich den Bedürfnissen der Wehrmacht, die in erster Linie Zweckmäßigkeit vorschreiben, anpassen. Auch er hat das feldgraue oder wüstenfarbige angelegt. Auf den ersten Blick scheint er dem

KdF-Wagen gar nicht ähnlich zu sein, aber dies ist nur äußerlich. In Konstruktion und Anordnung aller Teile ist er bis auf geringfügige Abänderungen der alte geblieben. Eine dieser Veränderungen ist die erhöhte Bodenfreiheit, die 29 Zentimeter beträgt gegenüber 22 Zentimetern beim KdF-Wagen. Diese Maßnahme trägt besonders dem unwegsamen Gelände Rechnung, er kann sich also nicht so leicht festfahren.

Die Einfachheit seiner Konstruktion hat sich an allen Fronten als ideal bewiesen, da sie eine schnelle und mühelose Auswechslung aller Einzelteile ermöglicht, falls dies einmal erforderlich sein sollte. Trotz der außergewöhnlichen Beanspruchung im Dienste der Wehrmacht genügt der kleine Ein-Liter-Viertaktmotor, der auch bei höchster Belastung des Wagens – 4 Mann mit Ausrüstung – nur 8 Liter Kraftstoff auf 100 Kilometer verbraucht. Von ganz besonderer Bedeutung im Einsatz in Afrika ist aber seine Luftkühlung. Sie macht den Wagen wirklich »unverwüstlich«, denn gleich, ob bei 40, 50 oder mehr Grad Hitze, er dreht unentwegt seine 3000 Touren, ohne sich heißzulaufen. Was es bedeutet, in der Wüste kein Wasser für den Wagen mitführen zu müssen, wird jeder, auch wenn er Afrika und die Gefahren der Wüste nicht kennt, beurteilen können.

Unser KdF-Wagen, der in den Zeiten des Friedens über 3 Millionen Versuchskilometer im verschiedensten Gelände zurücklegen mußte, ungeheuren Zerreißproben ausgesetzt wurde und dann erst nach restloser Bewährung für die Serienproduktion freigegeben wurde, hat sich als das beste Gebrauchsfahrzeug erwiesen. Eine Tatsache, die nicht nur den Konstrukteur, sondern vor allem auch die Sparer des KdF-Wagens mit Stolz und Beruhigung erfüllen kann, da sie so die absolute Gewißheit haben, daß ihnen bei Beendigung des Krieges, wenn die Serienproduktion für die zivilen KdF-Wagen aufgenommen wird, ein Fahrzeug ausgeliefert wird, das den letzten Erfahrungen der Automobilwirtschaft entspricht und das in seiner Haltbarkeit unerreicht ist. Dies ist aber auch die Grundbedingung für einen Wagen, der wirklich ein Volkswagen sein soll, denn hier muß mit dem Vermögen des einzelnen Volksgenossen gerechnet werden, der es sich nicht leisten kann, jedes Jahr seinen Wagen zu wechseln. Durch den Krieg ist die Produktion des KdF-Wagens zurückgestellt worden, da das Volkswagenwerk, wie jeder andere deutsche Betrieb, seine Arbeit auf vordringliche Bedürfnisse einstellen mußte. Überall, wo deutsche Truppen eingesetzt werden, auch der geländegängige Volkswagen mit dabei ist. Flink, zuverlässig und mit großer Wendigkeit erfüllt er seine Aufgaben. Das ist das Geheimnis unseres erfolgreichen Kampfes: wir haben nicht nur die besten Soldaten der Welt, sondern die unermüdlich schaffende Heimat stellt ihnen auch die besten Kampfmittel der Welt.«

Mit dem Ukas des »Generalbevollmächtigten für das Kraftfahrzeugwesen« Generalmajor von Schell vom 1. November 1941 wird die Produktion ›sämtlicher Sonderkonstruktionen der Wehrmacht auf dem Gebiet der Räder-Kfz.‹ ausgemerzt und durch den VW ersetzt. Somit verbleibt der Volkswagen mit seinen zahlreichen Abwandlungen im Typenprogramm der einzige PKW. Zur Aufnahme der verschiedenen Einbausätze des VW wird schon mit Order vom 2. 8. 1940 eine Verbreiterung des Aufbaues um 60–80 mm gefordert.

Unter der offiziellen Bezeichnung »leichter Personenkraftwagen (Kfz. 1) mit Fahrgestell des 1e Pkw K 1 Typ 82 Grundfahrzeug« liefen

folgende Abarten:
viersitziger Mannschaftswagen (Kfz. 1)
viersitziger Vermessungswagen (leichter Meßtruppkraftwagen – Kfz. 3)
dreisitziger Funkwagen (Funkkraftwagen – Kfz. 2)
zweisitziger Verwundeten-Transportwagen (für 1–2 Verwundete)
zweisitziger Sirenenwagen
zweisitzige Panzerkampfwagen-Nachbildung.
Auf sie ist immer Verlaß, denn wo die Geländefahrzeuge von Horch und Tatra nicht mehr anspringen, schnurrt der Porschemotor, kaum daß sich der Anlasser fünfmal gedreht hat, und erwärmt sich, wiederum dank seines raffinierten Kühlsystems, ohne Schwierigkeit auf günstige Betriebstemperatur.
Alle Ausführungen, im Volkswagenwerk von Wolfsburg gebaut, erhalten ab März 1943 (Motornummer 020000) den vergrößerten Motor mit 1131 ccm, während der ursprüngliche VW-Motor nur 998 ccm aufweist. Damit werden nun die vom Heeres-Waffen-Amt als unterste Leistungsgrenze für Militärfahrzeuge festgelegten 25 PS erreicht.
Die zahlreichen Kfz.-Instandsetzungsdienste übernehmen die technische Aufsicht der VW-Kübelwagen. Die erst im Kriege aus einer Improvisation heraus geschaffene Organisation der Kfz.-I.-Dienste erweist sich als überaus leistungsfähig und trägt viel zur Sicherstellung der Einsatzbereitschaft aller Truppen-Kraftfahrzeuge bei. Jedoch macht sich überraschenderweise ständig ein erheblicher Mangel an Kfz.-Schlossern unangenehm bemerkbar. Dabei stellt sich heraus, daß jene Kfz.-Handwerker sich am besten für die Truppe eignen, die ihren Zivilberuf in kleineren Ortschaften ausgeübt haben, weil sie es gewohnt sind, selbständig jede Situation zu meistern. Während sie nur zu oft in bespannten Einheiten oder an Stellen zu finden sind, wo ihre Kenntnisse überhaupt nicht verwendet werden können, muß man zahlreiche junge Rekruten in eigens dafür organisierten Schnellkursen notdürftig zu Kfz.-Schlossern ausbilden. Die schlechte Ersatzteillage bie allen Kfz.-I.-Diensten ist nur durch Erfahrung und Selbsthilfe einigermaßen auszugleichen. Die vielen, den ganzen Krieg hindurch nebeneinander laufenden Fahrzeugtypen erschweren die Beschaffung und Lagerhaltung von Ersatzteilen, und die starke Beanspruchung der Wagen unter feldmäßigen Bedingungen führt wiederum zu einem enormen Ersatzteilbedarf.

Am 23. Oktober 1942 gibt man im Reichsgesetzblatt die Stiftung *des Kraftfahrtbewährungs-Abzeichens* bekannt. Obgleich die Verordnung nur fünf kurze Artikel umfaßt, zeigt die im dritten Kriegsjahr erfolgte Stiftung, welche Anforderungen an alle diejenigen gestellt werden, die mit ihren Fahrzeugen im Einsatz sind. Nicht nur das Führen der Kraftfahrzeuge, sondern auch deren Pflege und Wartung gehören zu den Pflichten dieser Männer.
Das Abzeichen, als Anerkennung für den Kriegseinsatz besonders bewährter Kraftfahrer in den drei Stufen Gold, Silber und Bronze, wird in der Mitte des linken Unterärmels getragen.
Die Voraussetzungen für die Verleihung: »Einsatz ab 1. Dezember 1940 in den Gebieten des ehemaligen Jugoslawien, Griechenland, in Bulgarien, Rumänien, im Gebiet ostwärts der russischen Westgrenze 1940 (vor Eingliederung der Baltischen Staaten durch die Sowjetunion), in Finnland, Norwegen nördlich des Polarkreises oder im Gebiet der in Lappland eingesetzten deutschen Truppen und in Afrika.«

Die Vorgeschlagenen müssen sich unter besonders schwierigen Bedingungen bewährt haben: als Kradmelder an 90 Einsatztagen, als Fahrer von Gefechtskraftfahrzeugen an 120, als Fahrer von Kraftfahrzeugen der Trosse I und II, Kolonnen und Stäbe an 150, als Fahrer von Kraftfahrzeugen der Versorgungstruppen an 165, als Kraftfahrer sonstiger Dienststellen und Einheiten der Wehrmacht an 185 Einsatztagen. Als Einsatztage gelten bei Vorliegen erschwerter Unterbringungs- und Instandsetzungsverhältnisse z. B.: Fahrten unter Feindeinwirkung, besonders große Tagesleistungen an Strecke und Fahrzeit oder besonders schwierige Wegeverhältnisse oder Einsätze unter ungewöhnlich harten klimatischen Bedingungen.

Das Kraftfahrbewährungsabzeichen wird nicht nur an Wehrmachtsangehörige, sondern auch an der Wehrmacht unterstellte Nicht-Wehrmachtsangehörige, ferner an ausländische Freiwillige, sowie an »Freiwillige fremder Volksstämme aus den Ostgebieten – auch an entlassene Kriegsgefangene – soweit diese unter dem Befehl der Deutschen Wehrmacht im Rahmen bzw. in Verbänden der Deutschen Wehrmacht kämpfen« verliehen. Eine Verleihung an Wehrmachtsangehörige verbündeter oder befreundeter Länder ist nicht statthaft. Die Verleihung der nächsthöheren Stufe setzt jeweils die nochmalige Erfüllung der geforderten Bedingungen voraus.

Nord-Afrika, Herbst 1941: Eine Kraftfahrzeug-Instandsetzungswerkstatt des DAK. Die Vorderachse des an einem Hebegerüst aufgehängten VW-82 wird von dem Mechaniker überprüft.

Frankreich, Herbst 1941: Dr. Ley, Leiter der Deutschen Arbeitsfront (DAF),aus dessen Geheimfond das 120 Millionen RM Projekt des VW-Werks in Wolfsburg-Fallersleben finanziert wurde, startet zu einer Inspektionsreise entlang der Kanal- und Atlantikküste. Der werksneue VW-82 des Reichsministers hat eine extra Einrichtung: Der Kraftstoff-Einfüllstutzen ist mit einem Hängeschloß versehen.

Frankreich, Herbst 1941: Der VW als Rednertribüne. Dr. Ley spricht vor der versammelten Besatzung eines Kriegsmarine-Stützpunktes.

Frankreich, Herbst 1941: Der VW-Kübelwagen mit Reichsminister Dr. Ley trifft im Kriegsmarine-Stützpunkt Brest ein.

Frankreich, Herbst 1941: Nach der Besichtigung des Kriegshafens von Brest geht die Fahrt im VW-82 weiter.

Nord-Afrika, Winter 1941/42: Der VW-82/trop mit Sandreifen auf einer Wüstenstraße. Der hochgeklappte Deckel ermöglicht einen Blick in den Motorraum. Man sieht unter anderem die Kraftstoffpumpe, den Wirbel-Luftfilter und das Anschlußrohr für den Luftfilter. Unten das Abdeckblech mit Öffnung für die Andrehkurbel. Links unter dem Zughaken der Rückstrahler mit Pendel. Über dem Kotflügel das Abstand-Rückgerät Nova 1–2 A. Oben an der rechten Seite des Motorrahmens neben der Schlußleuchte das nachträglich eingebaute Haltegestell für den 20-Liter-Einheitskanister.

Nord-Afrika, Tripolis, Winter 1941/42: Feldmarschall der Luftwaffe, Albert Kesselring, auf einer Inspektionsreise mit dem VW-82, hinten.

Reichsgebiet, Winter 1941/42: Die Kraftfahrzeugkolonne einer motorisierten Einheit in Marschordnung am Rande einer Landstraße abgestellt.

Nord-Afrika, Libyen, Januar 1942: Links ein VW-82 der Feldpost. Rechts der »große Bruder«, der 8 t Zugkraftwagen vom Typ Krauss-Maffei KMm 11 (Sd. Kfz. 7).

Nord-Afrika, Winter 1942: Südlich von Port Brega ging dem »Weibele« die Luft aus. Der hintere Sandreifen des Kronprinzrades (200–12) wird aufgepumpt. Auf dem Vorderrad dagegen der Einheitsreifen (5,25–16).

Jugoslawien, Winter 1942: Eine Kraftfahrzeugkolonne im Partisanengebiet bei Sarajewo. Vorn ein VW-82 mit dem Ersatzrad im Schutzüberzug.

Über dem Mittelmeer, Frühjahr 1942: Eine recht unalltägliche Fahrt mit dem VW-82/trop im Stauraum einer Ju 52/3m. Der Kübelwagen ist mit Sandreifen Kronprinzrad (200-12) ausgestattet, die vier Insassen deutsche Afrika-Korps-Soldaten mit Schwimmwesten.

Nord-Afrika, Libyen, Frühjahr 1942: Ein VW-82 Kübelwagen vor einer wichtigen Straßenkreuzung in der Wüste unweit Maratuba. Rechts geht es, laut Wegweiser, unter anderem zu einem Stukageschwader, zu dem Gefechtsstand eines Regiments und zum Jagdgeschwader Waldenga.

Nord-Afrika, Libyen, 1942: Der VW-82/trop eines Nachrichtentrupps rollt mit Vollgas durch die Wüste.

5

Im Jahre 1941 läuft die Erprobung der VW-Modelle weiter. Der Bericht des *Ing. Herbert Kaes*, eines der Neffen des Prof. F. Porsche, und zugleich der Bruder von Porsches Privat-Sekretär, Ghislaine Kaes, vermittelt einen interessanten Einblick in eine der ausgedehntesten Versuchsfahrten während der ersten Halbzeit des Zweiten Weltkrieges: »Da die vorgesehene Erprobung der Fahrzeuge in Libyen durch Transportschwierigkeiten nicht unternommen werden konnte, startete die Kraftfahrversuchsstelle Kummersdorf in der Zeit vom 2. August bis 30. September 1941 zu einer Probefahrt durch das Protektorat, Ungarn, Jugoslawien, Bulgarien, Griechenland und Rumänien. Der für die Balkanerprobung vorgesehene KdF-Wagen nahm laut mündlicher Vereinbarung mit dem Heeres-Waffen-Amt WA Prüf 6 an der vom HWA veranstalteten Balkanerprobung von Rad- und Kettenfahrzeugen teil. Um Schwierigkeiten bei Grenzübertritten auszuschalten, wurde der Wagen mit einer Wehrmachts-Nummer WH-0204722 ausgerüstet. Der benötigte Brennstoff und das Öl stellte das HWA.

Die Kolonne bestand einschließlich der KdF-Limousine, zwei Kübelwagen Typ VW 87 und zwei vom Typ VW 82 aus 43 Radfahrzeugen. Der KdF-Wagen wurde auf allgemeine Tauglichkeit und etwaige Abänderung für den Verkauf in den Balkanstaaten erprobt.

Die Limousine, die auf der Versuchsfahrt insgesamt 12 986 km zurückgelegt hat, wurde zum größten Teil von Herbert Kaes gesteuert und etwa 3000 km von Fahrern der Vers.Kraft, Kummersdorf.

Erprobt wurden Pkw, Lkw und Motorräder sämtlicher im Generalmajor v. Schell-Programm vorgesehenen Radfahrzeuge. Mit einem Bahntransport kamen noch Kettenfahrzeuge (Kampfwagen, Zugmaschinen, NSU-Karetten) zum Bestimmungsort nach Griechenland.

Das Kommando über die Balkan-Erprobung hatte Hauptmann Ing. Rieger. Zur Beobachtung waren Ingenieure als Sonderführer kommandiert, von der Kraftfahrversuchsstelle eingesetzt. Jede Firma beteiligte Vertreter und Monteure. Die Monteure, zugleich Fahrer, sind zur Balkanfahrt vom HWA als Soldaten eingekleidet worden.

Für den einen Monat dauernden Aufenthalt in Griechenland schlugen wir ein Zeltlager bei Aggia-Marina an der Nordseite des Golfes von Lamia, 15 km von Lamia entfernt, auf. Die Zelte für die Beteiligten und Werkstätten waren teils Versuchszelte, teils englische Beutezelte. Wegen der großen Hitze spannten wir jeweils über das eigentliche Zelt mit einem Abstand von 20 cm Planen, um die starken Sonnenstrahlen abzuhalten. Für die Herren der Industrie stand ein großes Zelt zur Verfügung, in dem ich selbst nicht geschlafen habe. Zusammen mit dem Vertreter von Fa. Saurer, Herrn Krankenhagen, hauste ich in dessen 2-Mann-Zelt. Zwar lagen wir beide nicht wie die anderen in Feldbetten, sondern nur auf einer Decke am Boden, dafür aber allein und ungestört. Da wir nie bei Dunkelheit Licht machten, waren wir lange nicht so den lästigen Stichen der Moskitos und sonstigen Mücken ausgesetzt wie die anderen, die jeden Abend bis spät in die Nacht Licht im Zelt hatten. Verpflegung gab es aus eigener Feldküche, oder wir gingen in die nächste Ortschaft essen. Wasser zum Trinken und Waschen wurde täglich in Einheitskanistern herangeholt. Wir haben alle gerne die Unannehmlichkeit des Zeltlebens in Kauf genommen, denn mit den Hotel-Quartieren hatten wir sehr schlechte Erfahrungen gemacht. In den Balkanstaaten darf man nur in großen Städten und hier im besten Hotel übernachten, sonst hat man bestimmt Ungeziefer aller Art. Während des einmonatigen Aufenthaltes in Griechenland regnete es nur einmal kurz, aber ungemein heftig.

Straßenverhältnisse

Die Straßenverhältnisse in den Balkanstaaten sind zum größten Teil sehr schlecht: alte, seit Jahren nicht gerichtete Asphaltstraßen oder Staubwege mit nebeneinander liegenden Schlaglöchern und Steinen. Der Staub war so stark, daß der in Kolonne als 38. Fahrzeug hinter einem Panzerspähwagen laufende KdF-Wagen nur mit ständig eingeschaltetem Scheibenwischer fahren konnte. Die Fahrzeugabstände mußten bei der Kolonnenfahrt genau eingehalten werden. Auf der Fahrt nach und von Griechenland überwand die KdF-Limousine mehrere Bergstrecken, etwa wie die Glockner- oder Tauernstraße. Bei den Rundfahrten in Griechenland lief der Wagen allein sechsmal hin und zurück über den Thermopylenpaß, der eine Gesamtlänge von 25 km aufweist und etwa 12–15 % Steigung hat. Bei den verschiedenen Fahrten nach Athen, Delphi usw. fuhr das Fahrzeug nicht in Kolonne.

Streckenführung und Kilometerleistung

Anfahrt von Vers.Kraft Kummersdorf nach Griechenland über Protektorat – Ungarn – Jugoslawien – Bulgarien – zum Zeltlager bei Lamia	2 664 km
Rückfahrt von Zeltlager Lamia über Bulgarien – Rumänien – Ungarn – zur Vers.Kraft Kummersdorf	4 175 km
Rundfahrten in Griechenland	3 912 km
km-Stand des Fahrzeugs bei Abfahrt in Stuttgart	5 470 km
km-Stand des Fahrzeugs nach Beendigung der Balkanfahrt	16 221 km

Die KdF-Limousine hat sich im allgemeinen, trotz teilweise scharfen und hohen Fahrens sehr gut bewährt.
Verschiedene Herren der Industrie und des HWA, die kurze Strecken den Wagen selbst steuerten oder mitfuhren, waren voll des Lobes.

Brennstoff- und Ölverbrauch
Vergasereinstellung:
105 RD, 170 Luftzusatzdüse, 6 mm Düsenträ-

ger, 45 Leerlaufdüse, 21,5 ∅ Lufttrichter.
Brennstoffverbrauch bei Kolonnenfahrt auf guter Straße = 6,5 l/100 km
Brennstoffverbrauch bei Kolonnenfahrt auf schlechter Straße = 7,5 l/100 km
Brennstoffverbrauch nicht in Kol. gefahren auf guter Straße = 7,8 l/100 km
Brennstoffverbrauch nicht in Kol. gefahren auf schlechter Straße = 8,2 l/100 km
Der Wagen wurde zum größten Teil mit voller Belastung gefahren, 1050 kg Gesamtgewicht.
Mit dem in den Balkanstaaten getankten Brennstoff von 65–68 Oktan klingelt der Motor sehr stark. Nach Einbau einer 120 HD und eines 5,3 mm Düsenträgers war ein Klingeln nur bei stärkster Belastung der Maschine zu hören. Diese Einstellung war etwa 3000 km eingebaut.

Ölverbrauch
Bei Anfahrt – Griechenland = 0,3 Liter/1000 km
Bei Rückfahrt (bis Weidenbach ausgewertet, da hier Ölkühlerdichtung defekt wurde) = 0,75 Liter/1000 km

Motor
Verstaubung
Nach 4000 km Balkanfahrt Gummiabdichtung zwischen Motor und Aufbau stark mitgenommen, stellenweise eingerissen und aus Fassung gerissen. Motor verstaubt sehr.

Ansaugluftführung für Gebläse
Die serienmäßige Ausführung kann ohne weiteres beibehalten werden. Das Heranholen der gesamten Ansaugluft aus dem Wageninnern wurde nicht ausgeführt, da dies für die Serienausführung wegen des Motorgeräusches doch nicht in Frage kommen wird.

Filter
Der Einbau eines Ölsumpf-Filters ist unbedingt erforderlich. Bei Kolonnenfahrt im Staub nach 3000–3500 km Ölsumpf-Filter voll Staub. Fahrzeug allein gefahren, genügt alle 1000–1500 km Filteröl erneuern.
Da in den Balkanstaaten Kundendienstarbeiten recht mangelhaft und unregelmäßig ausgeführt werden, wäre der Anbau eines kleinen Filzbalgfilters zu empfehlen.

Ölverlust und Ölwechsel
Über 2/3 der Balkanfahrt Motor einwandfrei dicht. Ölverlust durch undichte Ölkühlerdichtungen bei 14 500 km-Stand des Fahrzeugs.

Saugrohr-Vorwärmung
Für Außentemperaturen von über 30° im Schatten ist, um Benzinpumpenstörungen zu verhindern, das Sperren der Saugrohrvorwärmung erforderlich. Bei Außentemperaturen unter 20° dagegen hat der Motor mit gesperrter Saugrohrvorwärmung schlechten Übergang. Zu empfehlen ist der Einbau einer regulierbaren Einstellung.

Verteiler
Stark verstaubt. Nach Aufsetzen einer Leinenschutzkappe einwandfrei abgedichtet.
Die Unterbrecherkontakte häufig angeschmort.

Zündkerzen
Bosch 175 T1 Zündkerze in Ordnung. Versuchsweise eingebaute 145 T7 Kerzen zeigten nach kurzer Zeit starken Elektroden-Abbrand.

Zündkabel
Die Kabel werden durch die Motorwärme brü-

chig und schlagen durch. Einigemale Zündstörung durch defektes Kabel. (Spule zu Verteiler.)

Lichtmaschine

Stark verstaubt. Zweimal mit Pinsel gereinigt. Störungen traten nicht auf.

Keilriemen

Kunstseide-Riemen nach 11 500 km Laufdauer gerissen.

Beim Fahren über die steilen und langen Bergstraßen im Balkan macht sich die Kraftreserve des großen Motors (1130 m^3) angenehm bemerkbar. Da die Motoren in den Werkstätten bestimmt nicht eine so gewissenhafte Pflege wie in Deutschland erhalten und in diesen Staaten alle Fahrzeuge weit überladen gefahren werden, ist der Einbau des verstärkten Motors zu empfehlen.

Federung

Wagen schlägt an Vorder- und Hinterachse auf. Nach Einbau eines 4- und 5-Blattfederstabs in die Vorderachse, sowie 27 mm ∅ Federstäben und doppeltwirkenden Stoßdämpfern in die Hinterachse Federung einwandfrei.

Bremse

Wirkung sehr gut. Am Ende der Versuchsfahrt zu weite Pedalluft. Bremse muß nachgestellt werden. Bremsbeläge gut.

Rahmen

Serienmäßiger Typ 82 Rahmen mit einfacher Verschweißung der Gewinderohre für Vorderachsbefestigung. Um ein vollständiges Ausreißen der Rohre zu verhindern, wurden bei km-Stand 9713 der Rahmenkopf seitlich aufgebohrt und die Rohre elektrisch an beiden Seiten verschweißt. Die Auflageplatte für Vorderachse wurde nachgerichtet und wieder autogen verschweißt.

Sonstiges am Fahrgestell

Lauffräder

Felgen stark verbeult.

Hintere Reifen nach 8000 km Gesamtlaufzeit Profil abgefahren. Reifen unbrauchbar.

Gegen Reifendurchschläge Luftdruck 1,6 atü hinten, 2,0 atü erhöhen.

Kraftstoffbehälter mit Leitung

Behälter wegen dauerndem Anschlagen des Hahn-Hebels an Spurstange etwa 5 cm höher gesetzt.

Brennstoffschlauch von Behälter aus Rahmen bei km 9145 eingerissen. Benzinbahn trotz Sieb über Rohre innen stark versandet. Störungen traten nicht auf.

Batterie

Nach 4000 km (seit Beginn der Versuchsfahrt) Batterie leer. Störungen traten nicht auf.

Aufbau

Verstaubung

Vorderer Kofferraum: Haubenabdichtung ungenau.

Motorraum: Gummiabdichtung zerstört.

Staubeindringen ins Wageninnere bei den Armaturen. Abhilfe durch Einlegen eines durchgehenden Pappestreifens vor den Armaturen. Zugleich Schutz der elektrischen Anlage.

Die Türabdichtung schlecht ausgeführt (Erster Serienaufbau)

Türen – Fenster

Die Türfenster ecken, wodurch bei gewaltsamem Kurbeln der Kurbelapparat verbogen

wird. Linkes Türfenster gebrochen, Zähne vom Zahnrad abgedrückt. Verbindungsstange von Türklinke verbogen.
Rechtes hinteres Seitenfenster aus Gummifassung gerutscht.

Sitze
Rücksitz hart.
Rückenlehne fällt dauernd samt Koffer nach vorn.
Abhilfe durch Anbringung zweier Riemen von rückwärtiger Aufbauwand zur Lehne.

Aufbaustoff
Der aufgeklebte Verkleidungsstoff löst sich von Aufbau.

Tachowelle
Gegen Beschädigung durch schwere Teile, wie Wagenheber usw. Welle im Kofferraum besser schützen.

Schlösser
Abschließbare Schlösser für Motor und vordere Deckelhaube im Balkan sehr zu empfehlen.

Kontrolle nach Versuchsfahrt
Äußere Lackierung stark angegriffen.

Verstaubung
Am Radeinbau und sonstigen Schweißfugen dringt Staub ins Wageninnere. Lenksäule innen stark verstaubt. Lenkradlager ausgeschmiergelt. Aufbau im linken hinteren Radeinbau mitte Strebenanschlag eingerissen.
Vorderteil unten durch Steinschlag verbeult.

Zuffenhausen
29. 11. 41 Ks/Kr.«

Von den vier VW-Kübelwagen, die neben der VW-Limousine auf der ›Balkan-Reise‹ getestet wurden, hatte man die beiden Fahrzeuge Typ VW 87, mit der Wagennummer 134 und 136, laut OKW-Auftrag vom 20. April 1941 für Tropenerprobung umgebaut. *Herbert Kaes* berichtet: »Die Fahrzeuge wurden auf Tauglichkeit in großer Hitze, Staub und Fahreigenschaft erprobt.

Streckenführung, Kilometerleistung und Straßenverhältnisse

In elf Fahrtagen von Kummersdorf über Prag – Wien – Budapest – Neusatz – Belgrad – Nisch – Sofia – Serres – Veria – Larissa – Lamis wurde der Bestimmungsort erreicht und am Golf, 15 km von Lamis, ein Zeltlager aufgeschlagen. Von hier nun wurden die Fahrzeuge teils auf schlechtesten Straßen oder ausgesuchten Staubrundstrecken im Gelände gefahren. Die Straßen in den Balkanstaaten sind als Zufahrtswege von den Militärfahrzeugen so stark mitgenommen, daß die in Kolonne fahrenden Wagen sich vollkommen einstauben. Auf der Hin- und Rückfahrt nach Griechenland mußten mehrere Bergstraßen, in Länge und Steigung der Großglocknerstraße gleich, überwunden werden. Die Kolonne bestand aus 43 Radfahrzeugen.

Kilometerleistung der Fahrzeuge

km-Stand bei Abfahrt	km-Stand bei Rückkehr Kummersdorf
Wagen 87/2 = 21 285 km	31 753 km
Wagen 87/4 = 34 727 km	38 441 km

gefahrene km
Wagen 87/2 = 10 468 km
Wagen 87/4 = 3 714 km (durch Unfall frühzeitig ausgeschieden)

Anfahrt nach Griechenland = 2600 km
Rückfahrt von Griechenland
Kummersdorf = 3400 km
Die Rückfahrt ging von Lamis aus über Veris – Serres – Sofia – Trnova – Bukarest – Weidenbach – Mühlbach – Großwardein – Budapest – Fürstenfeld – Gras – Holzkirchen – Hof – nach Kummersdorf.

Beobachtungen, Anstände und Abhilfe
Motor:
Kolben, Zylinder, Lagerabnützung

Nachdem die Fahrzeuge etwa 3000 km, davon 2600 km Anfahrt nach Griechenland, hinter sich hatten, wurden die Motore auf Sandeintritt kontrolliert. Im Schwimmergehäuse des Vergasers, sowie in den Düsen zeigte sich starker Sandansatz. Das Saugrohr war dick belegt mit ölgetränktem Sand (vom Wirbelölfilter). Auch das Motoröl, welches rund vor 1000 km erneuert wurde, war sehr sandhaltig.
Abhilfe und Ergebnis: An Wagen 87/2 wurde die Frischluftzuführung aus seitlichem Kofferraum angebaut. Eine wesentliche Besserung konnte nicht festgestellt werden, auch nicht an dem mit gleicher Frischluft ausgerüstetem Serien Typ 82, Wagen 2432. Nach dieser Beobachtung wurde Wagen 87/2 mit einem Filzbalg-Trockenfilter, der am linken hinteren Kotflügel befestigt und an den Wirbelölfilter angeschlossen wurde, ausgerüstet. Das Ergebnis war überzeugend gut. Im Wirbelölfilter war nach 300 km Fahrt im stärksten Staub praktisch keine Verdickung des Öles festzustellen. Die Vergaserdüsen, sowie das Saugrohr ohne jede Sandspur. Selbstverständlich müssen die Klemmanschlüsse vom Filzfilter zum Wirbelölfilter und von diesem zum Vergaser einwandfrei dicht angeschlossen sein. Da der am linken hinteren Kotflügel befestigte Filzbalg-Trockenfilter direkt im Bereich des aufgewirbelten Staubes der Hinterräder liegt und eine Anbringung des Filters am vorderen Kotflügel aus Zuleitungsschwierigkeiten kaum in Frage kommt, wäre vorzuschlagen, den Filter im Kofferraum seitlich oder unter den hinteren Sitz unterzubringen. Der Filter würde bestimmt in kleinerer Ausführung als der erprobte seinen Zweck erfüllen. Beispielsweise in der Größe der Motorradfilter 250 × 150 mm von der Firma BMW.
Die Kolben- und Zylinder-Abnützungen schwanken zwischen 0,02 bis 0,03 mm. Die Haupt- und Pleuellager waren leicht riefig. Ein Vergleich zwischen den beiden Fahrzeugen, wobei Wagen 87/4 nur mit Wirbelölfilter und normaler Luftführung gefahren wurde, konnte nicht gemacht werden, da dieser Wagen bald wegen Unfall ausschied.
Die Ventile waren stark mit Sand belegt, haben aber durchweg noch gut getragen.

Verstaubung des Motors
Die Verstaubung des Motors konnte nach Anbau der Ansaugluftführung bedeutend vermindert werden. Leider ist die Gummiabdichtung zwischen Aufbau und Motorverblechung nach kurzer Zeit so zerstört, daß von unten viel Staub in den Motorraum eindringen kann. Die Motorhaube sollte auch besser abgedichtet werden. Um ein Verstopfen des Gebläses durch Papier, Putzwolle usw. zu verhindern, müßte ein Drahtgeflecht angebracht werden.
Beim späteren Zerlegen der Motore konnte festgestellt werden, daß die Kühlrippen vom Zylinder und Zylinderkopf aus dem Wagen mit Ansaugluftführung bedeutend weniger versandet waren als beim Motor mit normaler Luftfüh-

rung. Die mit Sand verklebten Kühlrippen sind nach Abheben des Gebläses leicht mit Hilfe eines Drahtes oder Schraubenziehers und Wasser zu reinigen.

Ölkühler

Häufig werden die Ölkühlerdichtungen undicht, wodurch hoher Ölverbrauch und starkes Verkleben des Ölkühlers, Zylinder und Zylinderkopf mit ölgetränktem Sand auftritt.

Vergaser, Brennstoffpumpe und Saugrohr

Die Abil-Dichtung zwischen Vergaser und Saugrohr schrumpft zusammen. Merklicher Leistungsabfall.

Beide Typ 87 hatten abgedichtete Drosselklappenachsen, die das Ausschlagen durch Sandeindringen verhinderten. Das Abschalten der Saugrohrvorwärmung ist für Temperaturen von + 35 bis 40° im Schatten unbedingt erforderlich, doch sollte das Abschalten mit Hilfe leicht verstellbarer Klappen möglich sein, um bei Temperaturunterschieden von tags etwa 35° im Schatten und nachts 15° Übergangsstörungen zu verhindern.

Brennstoffpumpenstörungen treten erst bei Ölsumpftemperaturen von 110–115° auf. Um diese Temperaturen zu erreichen, mußte mit unserem Wagen vollbeladen eine Bergstraße, in Länge und Steigung der Großglocknerstraße ähnlich, mit Vollgas befahren werden.

Bei Betätigung der Hilfspumpe war die Störung sofort behoben.

Elektrische Anlage am Motor

Zündkerzen: starker Abbrand der Elektroden, mit Bosch-175 T1-Kerzen nach 1200 km 0,8 m/m.

Die Entstör-Kerzenstecker Bosch EM/W/10/40 halten nicht auf den Zündkerzen mit kurzem Gewinde.

Der Magneton-Magnet hat in der Anlaßdrehzahl schlechte Zündfunken, so daß das Anlassen recht erschwert war. Mit Batteriezündung springen die Motore auch besser wie mit dem Bosch-Magnet an.

Die Lichtmaschinen waren sehr stark verstaubt, doch traten keine Störungen auf.

Sonstiges

Ölverluste:

an oberer Gehäusetrennfuge;

an Gehäusebohrung für Magnetbefestigung; durch Einlegen eines Gummidichtringes behoben,

an Lichtmaschinenauflage durch Beilegen einer Papierdichtung in Ordnung.

Einbau von Motor-Temperaturmesser:

die eingebauten Fernthermometer für Ölsumpf haben sich für die große Hitze als äußerst notwendig erwiesen. Der Fahrer sieht von Zeit zu Zeit auf den Temperaturanzeiger und läßt den Motor nicht überhitzen.

Motorklingeln:

die Motore haben alle, solange sie mit dem in Griechenland getankten Brennstoff von 65–68 Oktan liefen, stark geklingelt.

Getriebe

Das Fünfgang-Getriebe bewährt sich sehr gut, doch sollte der Geländegang ohne vorheriges Ausschalten des Vierganggetriebes einschaltbar sein. Die Fahrer vergessen dies und verbiegen dann mit Gewalt das Hebelwerk.

Bei 6000 km Fahrtleistung von Wagen 87/2 fraß der vierte Gang auf Triebling.

Bei Kontrolle des Unfallwagens 87/4 war die

Büchse vom Rückwärtsgang ausgewalzt und der Bolzen angerieben.
Da das Getriebe verhältnismäßig wenig Öl faßt und durch den alten Entlüfter, sowie häufig undichten Hinterachssimmeringen viel Öl verliert, muß alle 2–3000 km das Getriebeöl nachgefüllt werden.

Hinterachse
Das Z. F., wie auch das Rheinmetall-Sperrdifferential, war nach der Balkanfahrt stark abgenützt.
Die Befestigungsmutter der Bremstrommel lockert sich trotz gutem Sichern. Bei Kontrolle konnte festgestellt werden, daß die Laufradwelle mit aufgekeiltem Zahnrad (alte Ausführung) durch die lockere Bremstrommel stark angeschlagen war (Nutenprofil). Die Hinterachssimmerringe werden schnell undicht, was starken Getriebeölverlust, sowie bei verstopftem Ablaufrohr vom Schleuderring ein Verölen der Bremse herbeiführt.
Die Gleitsteine mit Gelenkwellenfaust waren an beiden Fahrzeugen leicht angerieben. Schlechte Schmierung des Differentials.

Vorderradantrieb
Bei Zerlegen des Vorderradantriebs von Wagen 87/4 nach Balkanfahrt zeigten der Triebling und das Tellerrad leichte Freßstellen an den Zähnen. Das Öl war etwas mit Wasser durchsetzt.

Vorderachse
Lagerungen: An den Traghebel-, Achsschenkel- und Rundbolzenlagern traten keine Störungen auf. Lediglich die Bundbolzen wurden auf der gesamten Fahrt je zweimal nachgestellt.
Abdichtungen: Der Simmering wird nach kurzer Zeit undicht, was ein Verölen der Vorderradbremse mit sich bringt. Abhilfe durch bessere Abdichtungen und Einlegen eines Ölfangbleches, ähnlich wie an Hinterradbremse.
Die Ledermanschette gab keinen Anlaß zu Beanstandungen. Die Anschlaggummi für Vorderachse wurden durchgeschlagen und zur Seite gedrückt. Mit dem Einbau der neuen runden Anschläge wird dieser Fehler behoben sein.

Lenkung
An beiden Fahrzeugen brach nach einer Fahrtleistung von etwa 3500 km der Lenkhebel an der Lenkung. Die Bruchstelle war beide Male an der ersten Kerbe vom Befestigungsauge des Lenkhebels. Durch Verstärkung des Lenkhebels am Auge dürfte der Fehler behoben sein.
Die nachträglich eingebaute Begrenzung des Federweges zur Lenkmutter hat sich gut bewährt.
Durch grobes Fahren mit Wagen 87/2 im Gelände hatten die Vorderräder 1 cm Nachspur. Ein Verdrehen der Lenkung auf dem Achsrohr konnte nicht festgestellt werden. Folglich müssen die Lenkschenkel nachgegeben haben.

Federung
Beide Fahrzeuge hatten vorn je einen Drei- und einen Vierblattfederstab, hinten 29 mm Federstäbe.
Bei voller Beladung schwingt der Wagen in der Vorderfederung und schlägt manchmal an der Hinterachse auf. Durch Einbau von härteren Stoßdämpfern in die Vorderachse und 29 mm ∅-Federstäbe in die Hinterachse dürfte die Federung in Ordnung sein.
Die Gummitüllen der vorderen Stoßdämpfer werden ausgeschlagen.

Bremse
Solange kein Verölen durch undichte Simmeringe eintritt, ist die Bremse einwandfrei.

Laufräder
Der Reifendruck muß vorn auf mindestens 1,6 atü und hinten 2,2 atü erhöht werden, um ein Durchschlagen der Reifen zu verhindern.

Rahmen
Die Bodenbleche werden beim Fahren in schwerem Gelände durchgebeult.

Aufbau mit elektrischer Anlage
Die Aufbauten wurden weiter nicht beobachtet, da sie alte Ausführung sind und für die Fahrt nur repariert wurden.
Die Warnsummer sollten erst bei niederer Motordrehzahl einschalten. Die beiden Typ 87 werden laut Hauptauftrag Nr. 88 für Versuche in strenger Kälte umgebaut.

Zuffenhausen, 12. 11. 1941
gez. Kaes H.«

Griechenland, Mittelmeerküste, Frühjahr 1942: Ein VW-82 der Luftwaffe auf der Straße nach Elleniko. Der weiße Buchstabe V auf der linken vorderen Tür steht für Viktoria.

Nord-Afrika, Cyrenaika, Frühjahr 1942: Zwei VW-82-Besatzungen im Gespräch mit italienischen Kriegskameraden neben einer Wendemarke mitten in der Wüste. Das englische Orientierungsschild weist darauf hin, daß diese Gegend heiß umkämpft war.

Ostfront, Frühjahr 1942: Ein Korps-Kommandeur unterwegs. Der VW-82 trägt Wintertarnanstrich, vorn der Ständer mit dem Korps-Kommandoabzeichen.

Nord-Afrika, Tripolis, Frühjahr 1942: Die Besatzung eines VW-Kübelwagens ergänzt den Wasservorrat.

Nord-Afrika, Cyrenaika, Frühjahr 1942: Ein Offizier am Lenkrad eines VW-82 bei der Befehlserteilung. Im Hintergrund Kriegsgefangene und ein mittlerer Einheits-PKW.

Nord-Afrika, Libyen, Frühjahr 1942: Eine Patrouille in einem VW-82/trop (Sandreifen) mit MG 34.

Nord-Afrika, Libyen, Frühjahr 1942: Eine kurze Rast an der Via Balbia, der einzigen asphaltierten Straße der Cyrenaika. Neben dem VW-82 steht das ausgebaute Vordersitzgestell. Auf der Straße der Bus des Stabes eines Korps-Kommandos.

◄ Ostfront, Frühjahr 1942: Ein VW-82 der Panzergruppe 1 (v. Kleist). Der Scheinwerfer trägt eine Abblendkappe. Die umgeklappte Windschutzscheibe ist mit einer Schutzhülle überzogen. In der Karabinerhalterung auf der Anhaltestange hinter dem Fahrer ein K 98 b im Schutzüberzug.

Nord-Afrika, Libyen, Frühjahr 1942: In einem VW-82/trop der 2. Fallschirmjäger-Division beim Durchqueren der Senke bei Misurata. Links beide Steckfenster. ►

Inmitten der Senke von Misurata. ►

Nord-Afrika, Libyen, Frühjahr 1942: Bei einer Rast in der Wüste nahe Homs. Am Türgriff des VW-82/trop hängt eine MP-38/40. ▼

Die Senke von Misurata – eine Wasserwüste in der Wüste – scheint ohne Ende zu sein . . .

. . . geschafft! Wieder auf dem Trockenen.

Sizilien, Frühjahr 1942: Die Besatzung eines VW-Kübelwagens der Luftwaffe hat kurz Halt gemacht, um dieses Foto für das Familienalbum zu schießen.

Mittelmeerraum, Frühjahr 1942: Der Kübelwagen einer Luftwaffen-Propaganda-Kompanie (PK).

Ostfront Frühjahr 1942: Ein Kübelwagen durchfährt eine riesige Wasserpfütze auf einem Feldweg.

Ostfront, Frühjahr 1942: Der Kfz-Mechaniker einer Instandsetzungswerkstatt. Der hochgeklappte Deckel des Motorraums gibt den Blick auf den Motor frei.

Ostfront, in der Nähe von Lida, Frühjahr 1942: Die Kraftfahrzeug-Feldwerkstatt einer Luftwaffen-Einheit. Die Vorderachse eines VW-82 wird zur Reparatur abmontiert.

Ostfront, Frühjahr 1942: Hier war ein Dichter am Werk. Eine gereimte Warntafel für Kfz-Fahrer auf dem Markt von Kiew.

Ostfront, Frühjahr 1942: Eine willkommene Ergänzung der Wehrmachts-Speisekarte. Der Beifahrer des VW-82 beim Handeln mit den kleinen Eierverkäufern.

Nord-Afrika, Ende Mai 1942: Es geht vorwärts. Ein VW-Kübelwagen der Feldpost führt die kilometerlange Kolonne an.

Nord-Afrika, Cyrenaika, Juni 1942: Motorisierte Wachtposten eines KG-Sammel-Lagers in der Wüste.

Nord-Afrika, Tripolitanien, Juni 1942: »Hausputz« in einem VW-Kübelwagen, unweit Homs.

Eine motorisierte Heeres-Einheit in Marschordnung auf einer Landstraße in Frankreich. Vorn ein VW-Kübelwagen, hinten die leichte Halbketten-Zugmaschine 3 t (Sd. Kfz. 11).

Süd-Italien, Sommer 1942: Ein VW-Kübelwagen wird in ▶ ein Großraumtransportflugzeug geschoben.

Mittelmeerraum, Sommer 1942: Ein VW-82 wird für die Reise nach Nord-Afrika in einen Großraum-Segler vom Typ Gotha GO 242 A-1 verladen. Das große V bedeutet Viktoria.

Reichsgebiet, Sommer 1942: Der »Kriegskäfer« VW-87. Ein Versuchsmodell im Porsche-Werk Stuttgart-Zuffenhausen. Die mit einem 25 PS starken Motor ausgestattete VW-Limousine hat Allrad-Antrieb. Der VW-87, von dem nur 564 Fahrzeuge gebaut wurden, ging meistens zu den Rommel-Truppen nach Nord-Afrika.

Nord-Afrika, Sommer 1942: Eine Kfz-Feldwerkstatt in der Libyschen Wüste nahe Derna. Im Hintergrund ein VW-82 der Luftwaffe mit dem Zeichen der 21. Panzer-Division (DAK), der einer leichten Flak-Einheit angehört.

Nord-Afrika, Libyen, Sommer 1942: Die Besatzung des VW-82/trop als Selbstversorger. Vorn, aus dem Ersatzrad ragt die Drahtstange für den Erkennungswimpel. Die Windschutzscheibe ist umgeklappt und in einer Schutzhülle, der Rückblickspiegel umgedreht – eine Vorsichtsmaßnahme, um nicht durch Sonnenspiegelung dem Feind den Standort zu verraten.

Nord-Afrika, Cyrenaika, 4. August 1942: Die während der Kämpfe in der Gegend der Quathra Senke gefallene Besatzung eines VW-82.

Nord-Afrika, Cyrenaika, 4. August 1942: Die Trümmer einiger Fahrzeuge des 382. Grenadierregiments in der heiß umkämpften Gegend der Wüstenhaltestelle Tel-el-Eisa bei El Alamein.

Ostfront, Sommer 1942: Ein VW-82 einer Luftwaffen-Einheit.

Ostfront, Sommer 1942: Der fahrbereite VW-82 einer Heeres-Einheit.

Ostfront, Sommer 1942: Das vorläufige Ende einer Dienstfahrt. Der VW-82 rutschte vom schmalen Weg des Dammes in den Teich.

Mittelmeerraum, Sommer 1942: Ein im schweren Gelände abseits der Straße steckengebliebener VW-Kübelwagen.

Reichsgebiet, Sommer 1942: Der VW-82 eines höheren Luftwaffen-Stabes.

Nord-Afrika, Sommer 1942: Vor der Morgentoilette. Auf dem Kasten-Vorderteil ein zweckentfremdeter »Erste Hilfe«-Kasten mit persönlichen Utensilien. Das Mittelteilblech mit dem Nummernschild ist eingedrückt.

Sommer 1942: Ein VW-82 kollidierte mit dem Geländer einer italienischen Grenzbrücke. Ein verbeulter Kotflügel und das verlorene linke Hinterrad sind das Ergebnis.

6

Ende Dezember 1941 sind die Ergebnisse der Balkan-Fahrten ausgewertet und nur durch die Erfahrungen mit dem VW-Kübelwagen in Nord-Afrika zu ergänzen. So entsteht im Frühjahr 1942 ein für die Afrika-Korpsfahrer unentbehrliches Merkblatt über die Probleme des Wüsteneinsatzes der Kraftfahrzeuge:

Allgemeines
Die hohen Temperaturen, die großen Unterschiede der Tag- und Nachttemperaturen sowie die starke Sonnenbestrahlung wirken sich in den Subtropen und Tropen auf die Kfz. sehr nachteilig aus. Nur eine umfassende Kenntnis der nachteiligen Auswirkungen ermöglicht die Anwendung der jeweils wirksamsten Gegenmaßnahmen.

Motor
Zylinder, Gehäuse, Kurbeltrieb und Schmierung
Infolge der auch in der kalten Jahreszeit für den Kfz.-Betrieb günstigen Temperaturen, die kaum unter 0°C liegen, ist mit Schwierigkeiten beim Anlassen des Motors nicht zu rechnen. Dagegen ist die Gefahr der Überhitzung des Motors sehr groß. Wenn unter dem Einfluß der hohen Außentemperaturen das Schmieröl sehr dünnflüssig wird, neigen alle gleitenden Teile zum Fressen. An den Dichtungsstellen des Motors kann Öl nach außen treten. Hier bildet es mit dem in erheblicher Menge in der Luft befindlichen Staub und Flugsand eine dicke Isolierschicht, die den Abfluß der Motorwärme behindert, was zu weiterer Überhitzung und deren Folgen führt.

Kraftstoff
Ottokraftstoff neigt zur Bildung von Dampfblasen, die, wenn sie sich zwischen Kraftstoffbehälter und -pumpen befinden, zum Stehenbleiben des Motors führen. Durch die hohen Temperaturen ist ein Verdunstungsverlust in den Kraftstoffbehältern der Kfz. und in den Lagergefäßen bedingt, was bei der Mitnahme und dem Nachschub von Ottokraftstoff mengenmäßig zu berücksichtigen ist. Der Verdunstungsverlust ist je nach Jahreszeit verschieden.

Gemischbildung und Zündung
Infolge der hohen Temperaturen ist das spez. Gewicht der Luft kleiner. Der Motor erhält demnach ein geringeres Füllungsgewicht, erzeugt also unter sonst gleichen Bedingungen eine geringere Leistung als normal, wodurch ein höherer Kraftstoffverbrauch eintritt. Der Kraftstoffverbrauch kann durch Einsetzen einer kleineren Kraftstoffdüse verringert werden, doch ist zu prüfen, ob dann die Leistung des

Motors noch ausreicht. Durch die Kondenswasserbildung während der Nacht setzen sich Wassertropfen im Verteiler oder an den Elektroden der Zündkerze ab und verursachen Zündstörungen.

Elektrische Ausrüstung
Durch Verdunsten tritt bei den Sammlern ein hoher Wasserverlust ein. Hierdurch wird die elektrochemische Leistung des Sammlers stark herabgesetzt und die Betriebssicherheit gefährdet. Unter dem Einfluß des hohen Salzgehalts der Seeluft in Küstengebieten korrodieren die Klemmen der Sammler, was zu Störungen in der Stromzuführung führt. Bei hohen Temperaturen besteht Neigung zu Überladung.

Magnet- und Sammlerzündung
Hier führt das sich nachts bildende Kondenswasser zu Störungen und Versagen der Zündung (Unterbrecher und Verteiler).

Triebwerk, Laufwerk, Bremsen und Zentralschmierung
Bei richtiger Wartung kein betriebsstörender Einfluß durch höhere Temperaturen.

Aufbauten
Bei ungeschützt aufgestellten Kfz werden die Metallteile so heiß, daß ein Berühren leichte Verbrennungen verursachen kann. Teile aus Holz, Leder, Gummi, Textil- oder Kunstharzstoffen werden rissig, spröde und zerstört. Dazu kommt, daß die schnellen und starken Temperaturschwankungen Spannungen hervorrufen, die eine vorzeitige Zerstörung herbeiführen.

Bereifung
Hier gilt sinngemäß das zuvor Gesagte, jedoch ist besonders auf die starke Bodenerwärmung während des Tages und deren Einfluß auf den Luftdruck und die Lebensdauer der Bereifung hinzuweisen. Die Bodentemperatur, die 65 bis 80° C annehmen kann, erhöht naturgemäß auch die Temperatur der Bereifung stärker, als es schon unter normalen Verhältnissen durch die Walkarbeit der Fall ist; bei Luftreifen kann der Reifendruck bis auf das 1½fache des Anfangsdruckes steigen. Wird hierauf nicht geachtet und in den heißesten Tagesstunden bzw. zu schnell gefahren, so löst sich schon nach ganz kurzer Zeit die Gummiauflage von dem Gewebe der Luftreifen, bzw. bei Bandagen, Vollgummireifen und Gummipolster treten Zerstörungen durch Überhitzung auf. Reifen sollen auf den vorgeschriebenen Druck aufgepumpt werden. Sie dürfen in keinem Fall einen geringeren Druck haben, da sonst Schäden durch Wandern des Reifens auf der Felge und durch übermäßige Walkarbeit im Reifen selbst auftreten. Häufige Prüfung am Tage ist daher unerläßlich, ebenso guter Luftdruckmesser und gute Luftpumpe.

Allgemeines
Die Straßenverhältnisse in den Tropen können nicht mit denen Europas verglichen werden, sondern es ist zu berücksichtigen, daß es wirklich gute Straßen sehr selten und meist in nur beschränkter Länge gibt. Alle anderen sogenannten Straßen, Pisten und Karawanenwege sind lediglich durch die Natur oder durch Spuren gekennzeichnete Verbindungen zwischen zwei Punkten und stellen an das Kfz. und den Fahrer die gleichen Ansprüche wie reines Geländefahren. Die Tragfähigkeit der Bodenoberfläche ist vom Fahrzeug aus oft schwer zu

beurteilen und kann rasch wechseln. Dadurch wird eine erhöhte Aufmerksamkeit der Fahrer notwendig. Jedes Gelände, in dem Hindernisse auftreten oder vermutet werden, ist nach Möglichkeit zu erkunden und Umgehungswege zur Vermeidung der Hindernisse festzulegen. Für die Erkundung werden am besten Geländeerkunder in leichten Fahrzeugen eingesetzt, die auch die Kennzeichnung von Umgehungswegen vornehmen.

Geländearten

Sandwüsten
Diese bestehen in der Hauptsache aus einer mehr oder weniger starken Schicht grobkörnigen Sandes, der eine verhältnismäßig feste Oberfläche bildet, so daß für in gleichmäßig zügiger Fahrt befindliche Räder-Kfz., außer Krädern, kaum Schwierigkeiten auftreten, während sie für Ketten-Kfz. überhaupt nicht zu erwarten sind. Schwieriger wird das Fahren, wenn weiche Sandeinlagerungen oder Flugsanddünen zu überwinden sind, da hier in vielen Fällen ein Festfahren und Einsinken der Räder-Kfz. eintritt. Der Versuch, festsitzende Kfz. wieder freizumachen, wird, wenn nicht die richtigen Mittel angewandt werden, nie einen Erfolg haben, in vielen Fällen aber meist zum Bruch der Triebwerke führen.

Steinwüsten
Hier ist der Untergrund fest, die Oberfläche mit Steinen bis zu Kopfgröße bedeckt. Unterliegt die Gegend dem Einfluß großer Temperaturunterschiede, dann ist das Gestein gesplittert und sehr scharfkantig, so daß die Bereifung sehr beansprucht wird. In den Steinwüsten werden die Kfz. durch Verwindungen und Stöße übermäßig beansprucht.

Niedrige Sandhügel, die um Sträucher (Kameldorn) zusammengeweht wurden und unter dem Namen »Hummochs« bekannt sind.

Wadis oder Rullahs (ausgetrocknete Flußbetten)
Ihre Ufer sind oft sehr steil, und die Betten selbst bestehen aus weichem Sand. Eine Übergangsstelle ist zu erkunden und zu kennzeichnen. Oft kann durch geringe Planierungsarbeiten mit dem Spaten das Überqueren eines Wadis erleichtert werden.

Bewässerungsgräben, bei deren Überquerung Fahrzeuge leicht aufsitzen können.

Fester Sand
Dieser eignet sich vorzüglich zum Befahren mit Kfz., doch sind in ihm eingelagerte sumpfige Stellen zu vermeiden. Nasser Sandboden ist unter Umständen gut befahrbar.

Mit Gestrüpp bewachsener Boden
Im allgemeinen gut befahrbar, jedoch sind kesselartige Senken zu vermeiden, die an dem helleren Grün des sie umgebenden Buschwerks zu erkennen sind.

Kieswüsten
Im allgemeinen gut befahrbar. Oft findet sich hier über weite Strecken Sandboden, der an der Oberfläche eine Schicht aus Kie elsteinen trägt. Auf eingelagerte Stellen und Flugsand achten!

Salzpfannen
Durch Regenfälle ändert sich ihre Oberflächenbeschaffenheit sehr stark. In trockenem Zustand kann die Oberfläche rissig und borkig, aber auch staubig sein und aus einer Salzkru-

ste bestehen. Vorsicht ist geboten, da unter einer anscheinend trockenen Oberfläche bisweilen der Untergrund naß ist. Dunkle Flächen sind auf ihre Tragfähigkeit zu prüfen. Bei Salzpfannen ist stets Vorsicht am Platze.

Sumpfstellen
Eine Sumpfstelle ist nach einem starken Regenfall niemals von einer Kolonne zu überqueren, ohne daß vorher ein einzelnes Kfz. zur Erkundung vorgeschickt ist oder ein Mann zu Fuß einen geeigneten Weg erkundet hat. Geländeketten bedeuten beim Überqueren sumpfigen Geländes eine wesentliche Erleichterung.

Sonstige Bodenverhältnisse
In der Mehrzahl muß mit Geländeverhältnissen gerechnet werden, wie sie auch in Deutschland z. B. im Dünengelände oder in unwegsamen Gebirgsgegenden bzw. Sumpf- und Moorgelände zu finden sind. Sie bieten dementsprechend auch die gleichen Schwierigkeiten. Die vorgenannten Bodenverhältnisse finden auch auf andere tropische Gegenden, soweit sie wüstenähnlichen Charakter haben, Anwendung.

Staub und dessen Auswirkung

Allgemeines
Der schlimmste Feind der Kfz. ist der Staub und in seiner gröberen Form der Sand. In Wüstengegenden ist der Staub so feinkörnig wie Mehl und dabei sehr scharfkantig. Er gelangt trotz vorhandener normaler Luftfilter und Dichtungen überall in das Innere der Maschinenteile und verursacht überall, wo gleitende Teile vorhanden sind, infolge des hohen Gehaltes an reinem Quarz und in Verbindung mit dem Schmieröl oder Fett einen hohen Verschleiß. Bereitet schon im normalen Fahrbetrieb der Staub große Schwierigkeiten, so gilt dies erst recht für die Zeit, in der Sandsturm herrscht. Je nach der Gegend dauert er Stunden oder mehrere Tage an; er führt ungeheure Mengen Sand mit, die alles einhüllen und gegen die nur sorgfältigste Vorbereitung Schutz bietet.

Motor
Der Motor ist am allermeisten der schädlichen Einwirkung des Staubes ausgesetzt. Der in den Motor gelangende Staub bewirkt schon nach kurzer Zeit einen derartigen Verschleiß aller gleitenden Teile, daß der Motor vollständig überholt werden muß.

Triebwerk
Im Wechselgetriebe und Achsantrieb bildet der eingedrungene Staub zusammen mit dem Schmiermittel eine sehr wirksame Schleifpaste, die ei en sehr schnellen Verschleiß der Zahnräder und Lager bewirkt.

Sonstige Feinde der Kfz.-Ausrüstung
Hier sind in erster Linie die Insekten zu nennen, von denen die Termiten besondere Teile, die aus Holz gefertigt sind, sowie Textilien zerstören. Auch ungeschütztes Leder wird von Insekten unbrauchbar gemacht. Luftfeuchtigkeit, Kondenswasser und Niederschläge lassen ungeschützte Eisenteile schnell rosten.
Vorbeugende Maßnahmen sind grundsätzlich schon in der Heimat vor dem Abtransport zu treffen, da in den in Frage kommenden Ländern in den seltensten Fällen die Möglichkeit besteht, weder das notwendige Material zu beschaffen noch die erforderlichen Arbeiten ordnungsgemäß durchzuführen. Werden die Kfz.nicht in der notwendigen Ausrüstung gelie-

fert, dann sind behelfsmäßig die im folgenden angegebenen Schutzmaßnahmen zu treffen:

Pflege

Allgemeines
Die örtlichen Bedingungen mit Staub, Hitze und Schlamm stellen viel höhere Ansprüche an ein Kfz. und verursachen einen erheblich schnelleren Verschleiß aller gleitenden Teile als in der Heimat. Deshalb ist eine sorgfältige Behandlung, Pflege sowie häufigeres Abschmieren und Ölwechsel erforderlich. Der Fahrer muß in der Lage sein, kleinere Schäden an seinem Kfz. ohne Inanspruchnahme des Instandsetzungsdienstes in Ordnung zu bringen. Alle Vorgesetzten sind verpflichtet, dafür Sorge zu tragen, daß jede sich bietende Gelegenheit zur ordnungsmäßigen Pflege der Kfz. ausgenutzt wird. Unnötige Fahrten sind zu vermeiden; stattfindende Fahrten sind restlos auszunutzen.
Anstrichschäden müssen bei großer Luftfeuchtigkeit während der Nacht und dadurch bedingter Rostbildung sofort und sorgfältig ausgebessert werden. Leder hat in trockenem und heißem Klima eine kurze Lebensdauer und muß durch ständige Behandlung mit Lederfett oder anderen Imprägnierungsmitteln frisch gehalten und nach Möglichkeit durch Überzüge vor direkter Sonnenbestrahlung geschützt werden. Die Pflege hat nach den Gerätbeschreibungen und Bedienungsanweisungen der betreffenden Kfz. zu erfolgen; die nachstehenden Hinweise sind außerdem noch besonders sorgfältig zu beachten.
80 % aller Motorenschäden sind auf mangelhafte Staubfilterung zurückzuführen! Das gesamte, mit Wartung und Instandsetzung beauftragte Personal (einschließlich der Beifahrer), ist zu gründlichster und häufigster Überprüfung und Reinigung der Filteranlagen zu erziehen. Jedem Kfz. sind zur Durchführung einer sachgemäßen und jederzeit möglichen Filterreinigung geeignete Gefäße beizugeben! Die Reinigung der Luftfilter ist in ihrer Häufigkeit durch keine Vorschrift begrenzt: sie kann bei schwerem Ghibli täglich mehrmals notwendig werden! Bei Ölbadluftfiltern ist zum Auffüllen nach Möglichkeit Altöl zu verwenden. Achtung! In Beute-Kfz. englischer und amerikanischer Herkunft sind gelegentlich sogenannte chemische Luftfilter eingebaut. Ihre Reinigung ist nicht nur zwecklos, sondern sie führt zu ihrer Zerstörung. Dasselbe gilt für den Versuch, sie mit Öl zu tränken. Luftfilter der genannten Art müssen nach Ablauf der auf ihnen vermerkten Streckenleistung beseitigt und – soweit vorhanden – durch neue gleichartige oder durch Luftfilter deutscher Herstellung von gleicher Größe ersetzt werden!

Bereifung

Allgemeines
Zur Schonung der Bereifung sind größere Märsche in erster Linie bei Nacht sowie in den frühen Morgen- und Abendstunden durchzuführen.
Die Bereifung (auch Ersatzbereifung) sind durch Abdecken vor Sonnenbestrahlung zu schützen.

Luftdruck
Die Reifen sind genau auf den vorgeschriebenen Druck aufzupumpen. Eine Herabminderung des Luftdrucks um höchstens 25 % darf nur in Ausnahmefällen beim Wiederflottmachen kurzzeitig vorgenommen werden. Der Reifendruck ist eine Stunde nach Fahrtbeginn

und dann in Abständen von je 4 Stunden zu prüfen. Ist der Druck durch Walkarbeit und Sonnenbestrahlung angestiegen, so ist keine Luft abzulassen, er ist zu erhöhen, wenn er unter den Normaldruck gesunken ist. Da Gummi unter dem Einfluß von trockenem und heißem Klima rasch brüchig wird, dürfen Ersatzschläuche auf keinen Fall in der Verpakkung, in der sie empfangen werden, aufbewahrt werden. Sie sind leicht aufgeblasen und mit Talkumpuder bestreut im Fahrzeug hängend mitzuführen. Entsprechende Halter sind anzubringen.
Regelmäßige Marschpausen sollen so oft wie möglich vorgesehen werden, wobei, wenn möglich, die Fahrzeuge in den Wind zu stellen und besonders bei heißem aber staubfreiem Wetter die Motorhauben zu öffnen sind. Wird wegen einer Störung an irgendeinem Fahrzeug eine Marschpause eingelegt, so ist der Führer der Einheit dafür verantwortlich, daß die Zeit nicht unnütz vertan wird, sondern daß die Fahrzeuge nachgesehen und etwaige Schäden abgestellt werden.

Ölwechsel bei neuen und überholten Motoren
Die Motoren sind mit dem für die Tropen bestimmten Motoren-Einheitsöl »Tp« zu füllen. Beute-Motorenöle englischer oder amerikanischer Herkunft erfüllen nach bisheriger Erfahrung die deutschen Bedingungen. Die Zeiten für Ölwechsel, auch für neue oder überholte Motoren, sind je nach Staubanfall bis auf die Hälfte der in den entsprechenden Vorschriften angegebenen Abstände zu verringern.

Fahranweisung
Das gute Fahren in der Wüste und ähnlichen Gegenden hängt von folgenden Bedingungen ab:

Der Fahrer muß über sein Fahrzeug genau Bescheid wissen.
Er muß imstande sein, jede Lage klar zu erkennen und die entsprechenden Maßnahmen zu treffen.
Er muß praktische Erfahrung im Fahren in der Wüste haben.
Er muß Kartenlesen können und die Anwendung des Marschkompasses beherrschen.
Jeder Fahrer muß dazu erzogen werden, vor Antritt der Fahrt den Bestand an Trinkwasser und Verpflegung ebenso zu überprüfen, wie dies von ihm in bezug auf Kraftstoff, Schmiermittel, Kühlwasser und Ersatzteile verlangt wird.
Keine Zeit, die auf Geländeerkundung verwendet wird, ist verschwendet. Jedes Gelände ist sorgfältig zu untersuchen und, wenn nötig, der Weg zu kennzeichnen.
Auf eingeborene Führer sich nicht verlassen, wenn es sich um große Strecken quer durch die flache Wüste handelt. Wenn diese Führer nicht besonders geschult sind, haben sie kein Gefühl für die ihnen ungewohnten Geschwindigkeiten und verlieren den Überblick. Ein zuverlässiger Dolmetscher dagegen ist sehr von Nutzen.
Märsche sind aus Tarnungsgründen sowie zur Schonung der Kfz. und Bereifung vorwiegend in Morgen- und Abendstunden oder bei Nacht (nur bei Straßenfahrten) durchzuführen.
Ein Gelände mit harter, glatter Oberfläche (z. B. trockene Salzseen) gestattet eine erhöhte Fahrgeschwindigkeit; jedoch ist niemals höhere Geschwindigkeit auf Kosten der Orientierung zu wählen, da sonst leicht ein Verlust an Zeit und Kraftstoff die Folge ist.
Beim Kolonnenfahren über Sandstrecken ohne festen Untergrund sind die Spuren voranfahrender Fahrzeuge, aber auch alte Spuren zu vermeiden. Bei ausgefahrenen Wegen mit felsi-

gem Untergrund ist darauf zu achten, daß Beschädigungen des Fahrwerks bzw. Laufwerks vermieden werden.
Sind Flugsand- oder Schlammstrecken erkannt, so ist der Gefahr des Festfahrens durch Auslegen von Unterlegblechen vorzubeugen.
Gut bewährt haben sich zusammenrollbare Unterlegbahnen (Rolleitern) aus Hanfseil mit Holzsprossen von doppelter Fahrzeuglänge, die leicht untergebracht werden können.
Gestaffeltes Fahren – oft in breiter Front – verringert die Beanspruchung der Fahrer durch den aufgewirbelten Staub vorausfahrender Fahrzeuge und ermöglicht ihnen damit gleichzeitig eine freiere Sicht. Bei Ausbruch eines Sandsturms sollen Kolonnen, sofern es die Gefechtslage irgend gestattet, unverzüglich anhalten und aufschließen. Die Motoren sind abzustellen und die Motorhauben zu schließen, Verdeck vom Pkw gut befestigen, Seitenteile aufstecken. Motorhaube abdecken. Wenn möglich, Motor mit Plane einhüllen. Auspuffrohr mit Kappe oder Stopfen verschließen. Nach Sandstürmen zuerst Motor reinigen.
Bei Rückfahrten nicht in der alten Spur fahren, sondern neben der Spur des zurückgelegten Weges entlangfahren.
Beim Abstellen von Kfz. mit überhitzten Motoren ist die Zündung niemals sofort auszuschalten, sondern die Motoren sind im erhöhten Leerlauf mindestens 5 Minuten weiterlaufen zu lassen.
Springen die Motoren morgens nicht an, so ist dies häufig auf Kondenswasserbildung im Verteiler oder an den Elektroden der Zündkerzen zurückzuführen.
Beim Marsch ohne Licht wenigstens das Abstandrücklicht des Nachtmarschgerätes einschalten, um Auffahren nachfolgender Fahrzeuge zu vermeiden.
Gleichmäßiges Fahren ist notwendig. Bei plötzlichem Gasgeben können die Räder mahlen, wodurch sich das Kfz. in den Sand eingräbt. Es ist möglichst geradeaus zu fahren, scharfe Kurven sind zu vermeiden. Beim Fahren mit Rückenwind ist öfter zu halten und der Kühler gegen den Wind zu stellen oder so schnell zu fahren, daß Fahrtwind zur Kühlung bleibt. Lüfterantriebsriemen prüfen.
Bei Kfz. mit zusätzlichem Frontantrieb ist dieser beim Überqueren von schwierigen Stellen (Sand, Schlamm und Geröll) einzuschalten.
Sollten sich nach dem Wiederauffahren auf eine feste Fahrbahn beim Ausschalten des Frontantriebs Schwierigkeiten ergeben, so ist dies auf vorübergehende Verspannung im Triebwerk zurückzuführen. Diese kann durch Fahren einiger leichter Kurven oder durch geringes Zurücksetzen des Kfz. ausgeglichen werden.
Beim Anhalten müssen die Bremsen so weich wie möglich betätigt werden. Geschieht es plötzlich, so gräbt sich häufig das Kfz. in dem weichen Sand ein.
Lufteintritts- und Austrittsöffnungen der Kühlanlage während der Fahrt stets freihalten. Keine Geräte usw. vor den Kühlern befestigen.
Geschwindigkeitsmesser mit Kilometerzähler und Uhr müssen in Ordnung sein.
Einheitsbehälter nur bis eine Handbreit unter Einfüllöffnung füllen. Gefüllte Kraftstoffbehälter, insbesondere Einheitsbehälter, so langsam öffnen, daß der Überdruck entweichen kann, ohne daß Kraftstoff herausspritzt.
Beim Veranschlagen des Kraftstoff- und Ölbedarfs für ausgedehnte Wüstenmärsche muß berücksichtigt werden, daß der Kraftstoffverbrauch wesentlich höher liegt als bei Straßenfahrten, so daß entsprechende Zuschläge gemacht werden müssen.

Auch im Sommer werden häufig bei Halten die Motoren im Leerlauf laufengelassen, unter Umständen sogar über längere Zeit. Abgesehen von der nutzlosen Verschwendung von Kraftstoff entstehen hierdurch schwerwiegende Schäden für den Motor, denn das mit Kraftstoff überfettete Leerlaufgemisch wäscht den Ölfilm an der Zylinderlaufbahn ab und verursacht durch fast trockene Reibung zwischen Kolben und Zylinder vorzeitigen Verschleiß. Ein Teil des Kraftstoffes gelangt ins Schmieröl, verdünnt dieses übermäßig und führt in der heißen Jahreszeit unvermeidlich zu Lagerschäden. Die Kerzen verkrusten und zünden nicht mehr.

Verschiedene Hindernisse und ihre Überwindung

Loser Sand
Schmale Streifen aus losem Sand werden in möglichst hohem Gang durchfahren, da hierdurch vermieden wird, daß sich die Räder durch Mahlen einwühlen. Wenn es sich herausstellt, daß ein Kfz nicht durchkommt und ein Herunterschalten auf einen niedrigen Gang notwendig wird, ist zu schalten, ehe der Motor von seiner Drehzahl heruntergekommen ist, oder das Kfz zu sehr an Fahrt verloren hat. Ist eine Sandfläche aber zu breit, als daß man sie mit Schwung durchfahren könnte, ist ein niedriger Gang zu wählen und die Motordrehzahl hochzuhalten. Ein Gangwechsel in weichem Sand ist zu vermeiden; nicht zu früh wieder heraufschalten.
Wenn die Tragfähigkeit des Sandes nicht mit Sicherheit erkannt werden kann, muß der Weg zu Fuß erkundet werden.

Wege in Ackerbaugebieten
Wege in Gebieten mit Ackerbau sind in der Regel in der Form von Dämmen längs der Kanäle und Bewässerungsgräben angelegt. Es ist darauf zu achten, daß die Straßenränder nicht befahren werden, da diese leicht unter dem Gewicht der Kfz einbrechen.

Geröll
Auf Geröllflächen mit großen Steinen ist darauf zu achten, daß nicht nur mit den Vorderrädern ausgewichen wird, sondern daß auch das Wechsel- oder Ausgleichsgetriebe und die Hinterräder nicht beschädigt werden.
Im ersten Gang und nur mit äußerster Vorsicht befahren. Höhere Geschwindigkeit nur, wenn es ohne Gefährdung des Fahrzeugs möglich ist.

Regenrinnen
Langsam und in niedrigem Gang fahren und sie durch schräges Anfahren überwinden, da sonst leicht Federbrüche eintreten.

Überqueren der Eisenbahn
Eine Eisenbahnlinie ist oft ein äußerst störendes Hindernis, besonders wenn sie auf einem niedrigen Eisenbahndamm verläuft. Bei Eisenbahndämmen muß die Möglichkeit der besten Auffahrt sorgfältig untersucht werden. Die Zeit, die für den Bau von Rampen verwendet wird, ist nie vergeblich. Zum Bau von Rampen können benutzt werden: Trittbretter, Sandbleche, Holzklötze, Sandsäcke, Steine und ähnliches.

Wadis, Gräben und Abhänge
Sind die genannten Geländearten schwierig, so ist langsam mit der nötigen Vorsicht heranzufahren; erst wenn der Fahrweg genau erkundet ist und etwa notwendige Wegverbesserungen durch Auslegen von Steinen oder Beseiti-

gung von Hindernissen vorgenommen sind, ist weiterzufahren. Wenn das Bett des Wadis aus Kies und Steinen besteht, so ist vorsichtig zu fahren und das Lenkrad fest in der Hand zu halten.
Das Überschreiten von Kammlinien muß vorsichtig erfolgen, da auf der anderen Seite ein steiler Abfall oder Überhang sein kann.
Beim Überwinden dieser Geländearten ist auf das führende Kfz. zu achten, um Zwischenfälle zu vermeiden.

Durchschleusen von Kfz. durch unbefahrbare Weg- oder Geländestrecken (tiefer Schlamm und Moor)
Beim Durchschleusen von Kfz. über schwierige Strecken dürfen mehrere Fahrzeuge durch Schleppseile nicht zu einem Schleppzug zusammengekuppelt werden. Es bringt in der Regel keine Zeitersparnis, sondern ruft nur schwerste Beschädigungen der Kfz. und der Schleppvorrichtung hervor. Die Fahrzeuge dürfen nur einzeln, höchstens zu zweien geschleppt werden.
Stehen an einer schwierigen Strecke eine Anzahl von Kfz., die geschleppt werden müssen, so sind vom Führer der Kolonne oder vom Dienstältesten sämtliche Mannschaften zusammenzufassen und zur Hilfeleistung anzusetzen.
Ein Teil hilft durch Schieben der jeweils geschleppten Kfz., der Rest gräbt die Straße frei und schafft durch Aufbringen von Steinen und kurzem Astwerk (nicht länger als ½ m) eine geeignete Fahrbahn. Die Äste sind quer zur Fahrtrichtung zu legen.
Das Legen von Baumstämmen und Bohlen in Längsrichtung der Spurrillen ist zu vermeiden, da diese leicht mit dem unbelasteten Ende hochgehen und Beschädigungen verursachen.

Aufstellen der Kfz.
Im allgemeinen werden die Kfz. im Freien meistens in Verbindung mit dem Zeltlager der Truppe aufgestellt. Grundsätzlich sind die Kfz. so aufzustellen, daß der Motor im Windschatten liegt. Die Öffnung der Auspuffleitung ist mit einem Stopfen oder einer Kappe zu schließen. Wenn möglich, ist die Motorhaube durch Abdecken gegen Flugsand zu schützen, wobei die Plane bis auf den Boden reichen muß und dort durch Sand oder Steine zu befestigen ist.
Durch Abdecken ist ebenfalls dafür zu sorgen, daß die Reifen nicht der direkten Sonnenbestrahlung ausgesetzt sind. Gefüllte Wasser- und Kraftstoffbehälter sind in den Schatten zu bringen bzw. unter dem Kfz. abzustellen. Bei längerem Aufenthalt sind die Kfz. zum Schutz gegen Sonne und Splitterwirkung in eine Grube zu fahren. Die Tiefe der Grube soll etwa 1 m betragen, so daß nach Möglichkeit die Oberkante der Reifen noch innerhalb der Grube selbst oder des sie umgebenden Erdwalles ist.
Auf Felsboden wird ähnlicher Schutz durch Aufschichten von Steinen erreicht. Ist Laubwerk oder ähnliches zur Verfügung, so ist auch das Kfz. selbst gegen den dauernden Einfluß der Sonne zu schützen, zumal durch das Abdecken mit Zweigen usw. gleichzeitig eine wirksame Tarnung gegen Fliegersicht erreicht wird.

Biwak
Für das Anlegen eines Biwaks gelten vom kraftfahrtechnischen Standpunkt aus folgende Gesichtspunkte:
Gute An- und Abfahrtmöglichkeit von der Straße oder Hauptversorgungspiste aus.
Leichte Tarnungsmöglichkeit, steinfreier Boden, damit die Kfz. mindestens bis zur Reifen-

oberkante eingegraben werden können.
Dem Anstrich der Kfz. entsprechender Untergrund.
Möglichst staubfreie Lage, Anfahrweg quer zur Hauptwindrichtung legen.
Kühler Platz, damit auch in den Mittagsstunden an den Kfz. gearbeitet werden kann.
Abstand von der Nachschubstraße mindestens 500 m.
Fahrzeugaufstellung vom Lagerplatz etwa 200 m abgesetzt.«
(Aus dem *Merkblatt des OKH für den Kfz.-Tropeneinsatz, Berlin 1942*)

Sommer 1942: Ein VW-82 ist an einem Hang abgerutscht. Der Fahrer, ein Kriegsberichterstatter, bei der Bergungsarbeit.

Sommer 1942: Der von der Bergstraße abgekommene VW-82 bietet den seltenen Blick auf das Unterteil des Fahrzeugs. Der Rahmen mit dem vorderen und hinteren Abdeckblech ist gut sichtbar.

Sommer 1942: Mit den am Vorderachs-Körper befestigten Stellen wird der VW-82 auf die Straße hochgezogen.

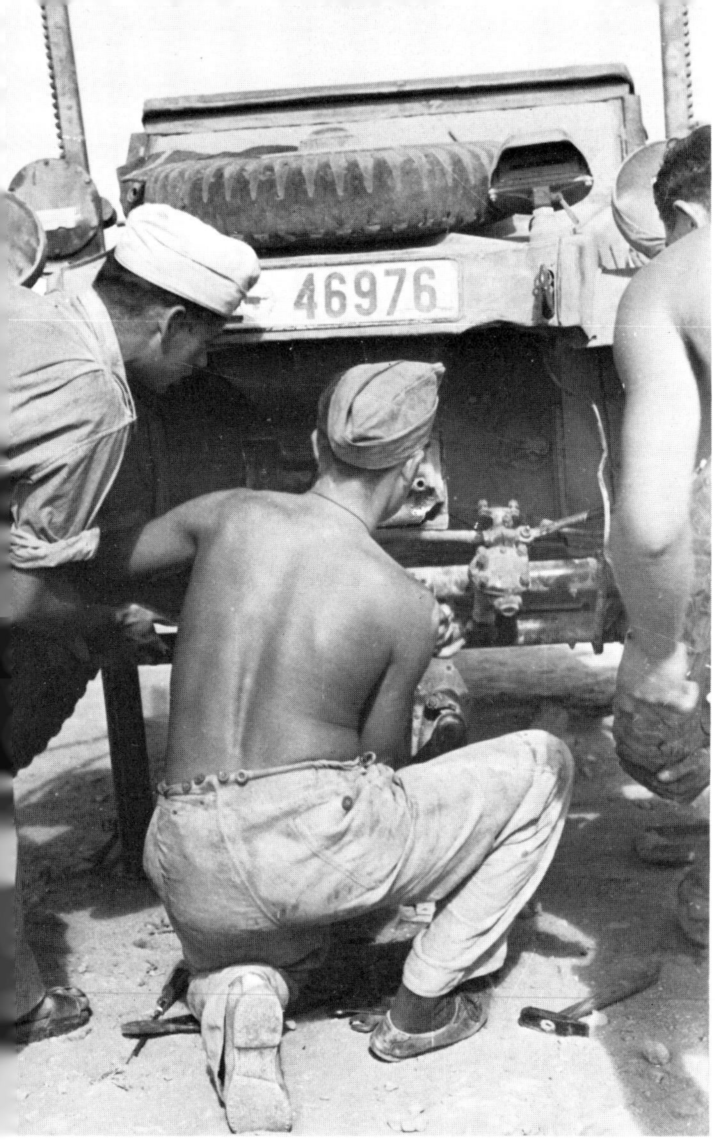

Nord-Afrika, Libyen, Sommer 1942: Eine Kfz-Instandsetzungswerkstatt in der Wüste. Der Kübelwagen einer Aufklärungsstaffel wird zur Vorderachsenüberholung aufgebockt. Vorn links das Lenkgehäuse.

Die improvisierte Hebebühne der Kfz-Instandsetzungswerkstatt in der Libyschen Wüste.

Gesamtansicht der Vorderachse des VW-82 Kübelwagens.

Nord-Afrika, Tripolis, Sommer 1942: Ein VW-82/trop auf der Via Balbia. Bei dem mit Kronprinz-Sandreifen ausgestatteten Fahrzeug hat man die vordere Tür wegen der Hitze abmontiert.

Nord-Afrika, Sommer 1942: In einer Oase am Rande der Libyschen Wüste. Zwei exotische Passagiere eines VW-82. Die geöffnete Tür gestattet einen Blick auf die beiden Vordersitze, bestehend aus dem Gestell und den Kissen.

Nord-Afrika, Libyen, 1942: Mitten in der Wüste bei sengender Mittagssonne, ein Schnappschuß für das Fotoalbum. Auf dem vorderen Kastenteil das Ersatz-Sandrad in einer Schutzhülle, oben darübergespannt das Fliegererkennungszeichen. Auf den Kotflügeln die Scheinwerfer mit Abblendkappen, rechts der Tarnscheinwerfer. Links neben dem Ersatzrad der Einfüllstutzen des Kraftstoffbehälters. Über dem linken Kotflügel die nachträglich angebrachte Halterung für den 20-Liter-Einheitskanister.

Nord-Afrika, Tripolitanien, Herbst 1942: Idylle in der Dzeebaa, eine durch Regenwassereinwirkung geschaffene Schlucht in der Wüste. Die Kübelwagen-Besatzung sucht Schutz vor Feindeinsicht. Auf der Windschutzscheibe eine Fußmatte aus dem Wageninnern, links der Suchscheinwerfer. Am vorderen Kastenteil das Ersatz-Sandrad mit einem Schutzüberzug. Vor dem Wagen das Vordersitzgestell.

Tripolis, Herbst 1942: Zwei Gegner posieren für ein Erinnerungsfoto. Auf dem Beifahrersitz des VW-82/trop ein soeben abgeschossener RAF-Pilot.

Nord-Afrika, Tripolis, Herbst 1942: Zwei VW-82 Kübelwagen/trop der 2. Fallschirmjäger-Division. Auf dem Hintersitz des vorderen Fahrzeugs ein einheimischer »Hiwi«.

Ostfront, Herbst 1942: Der ehemalige Dachauer KZ-Kommandant, Mörder des SA-Gründers Ernst Röhm und Erfinder des bürokratisierenden KZ-Terrors, SS-Obergruppenführer Theodor Eicke, zur Zeit Kommandeur der SS-Totenkopf-Division. Eicke fiel im Februar 1943 im Raum Charkow.

Nord-Afrika, Herbst 1942: In der Libyschen Wüste irgendwo zwischen Ras-Lanuf und Surt. Ein Kübelwagen der Luftwaffe neben einer Nachschubkolonne.

Nord-Afrika, Herbst 1942: Ein VW-82/trop mit Sandreifen hat Anlaßschwierigkeiten.

Tripolis, Herbst 1942: Ein mit allerlei possierlichen Exemplaren der Wüsten-Fauna geschmückter VW-82/trop des Kommandeurs einer der Infanterie-Einheiten des deutschen Afrika-Korps.

Nord-Afrika, Libyen, Herbst 1942: Ein in der Nähe von Tobruk liegengebliebener VW-82 des DAK.

Nord-Afrika, 13. November 1942: Ein britischer Konvoi am Halfaya-Paß. Dazwischen ein VW-82/trop.

Sizilien, Winter 1942/43: Der VW-Kübelwagen einer Propaganda-Kompanie des DAK auf dem Weg zu einem Heeres-Versuchsplatz.

Reichsgebiet, Schwarzwald, Winter 1942/43: Gelände-Versuchsfahrten der VW-82/Porsche Typ 155 Schneeraupe IV.

Reichsgebiet, Schwarzwald, Winter 1942/43: Die Gelände-Versuchsfahrten des VW-82/Porsche Typ 155 Schneeraupe IV.

Nord-Afrika, Tunis, Februar 1943: Ein Sanitäts VW-82/trop mit einem »Rote Kreuz« Fliegererkennungszeichen. Links neben dem Fahrer eine Doppeltragbahre für den Verwundetentransport. Im Hintergrund ein sorgfältig getarnter 12 t-Zugkraftwagen schwer (Sd. Kfz. 8) der Fa. Daimler-Benz.

Nord-Afrika, Tunis, Februar 1943: Der Kübelwagen VW-82/trop der 2. Fallschirmjäger-Division am Rande der Stadt Gabes.

Nord-Afrika, Tunis, Februar 1943: Ein VW-82 Kübelwagen der 2. Fallschirmjäger-Division auf einer Wüstenstraße in der Nähe des Kasserine-Passes.

Brückenkopf – Tunis, Februar 1943: Ein VW 82/trop der Luftwaffe mit einem gefangenen Tommy in der Nähe des Kasserine-Passes.

Ostfront, Februar 1943: Vorn rechts Generaloberst Hoth auf einer Frontinspektionsfahrt im VW-82, am Vortag der deutschen Gegenoffensive zwischen Dnjepr und Donez.

Nord-Afrika, Tunis, Mitte Februar 1943: Der VW-82/trop im unwegsamen Gelände des Kasserine-Passes. Einige Tage später führt Rommel hier erfolglose Verteidigungskämpfe gegen das II. US-Korps.

Nord-Afrika, Tunis, Mitte Februar 1943: Der VW-82/trop auf dem Wege von Bir-el-Afey zum Kasserine-Paß.

Der aufgeweichte Weg in der Nähe des Kasserine-Passes. Neben dem VW-82 ein 3 t Lkw Ford.

Der Kübelwagen 82/trop im schlammigen Gebiet des Kasserine-Passes.

Nicht nur an der Ostfront macht der Schlamm Mensch und Maschine zu schaffen. Dem VW-82/trop helfen selbst die Kronprinz-Räder mit den Sandreifen nicht viel.

Ostfront, Frühjahr 1943: Der Kübelwagen einer Panzer-Einheit hat sich auf dem aufgetauten Ackerfeld festgefahren.

Italien, Frühjahr 1943: Der VW-82 auf einer Bergstraße in den Apenninen.

Süd-Italien, Frühjahr 1943: Der Fahrer eines VW-82. Der Soldat trägt die Feldmütze des Deutschen Afrika-Korps.

Sizilien, Frühjahr 1943: Kurz nach dem Sonnenaufgang beginnt bei dieser Besatzung eines VW-Kübelwagen der Tag mit der Morgentoilette und Frühstück.

Italien, Frühjahr 1943: Im unwegsamen Gelände der Arno-Mündung westlich von Pisa. Der VW-82 arbeitet sich durch die schlammige Ebene . . .

. . . doch die Natur ist stärker. Der Kübelwagen mit der Nummer WH-922571, voll beladen mit Proviant für eine abgelegen stationierte Einheit, bleibt stecken.

Der WH-922571 ist mit der Hinterachse fast völlig eingesunken.

Schritt für Schritt wird der WH-922571 aus dem Morast geschoben . . .

... ein paar Zweige unter den Hinterrädern helfen weiter ...

... noch eine letzte Anstrengung, und der WH-922571 hat endlich wieder festen Boden unter den Rädern.

Ostfront, Frühjahr 1943: Ein VW-82 mit weiß getupftem Wintersichtschutz vor einer Pionierbrücke über den Dnjepr, südlich von Kiew.

Italien, Frühjahr 1943: Auf einer Landstraße in den Apenninen der VW-82 einer Nachrichten-Einheit.

Ostfront, Sommer 1943: Die linke Seite eines VW-82 Kübelwagens mit Verdeck und Steckfenstern.

Ostfront, Sommer 1943: Ein VW-82 unter einer Baumgruppe zum Schutz vor Fliegersicht.

VW-82 Kübelwagen: Auf der Fahrt durch weglose Steppe.

Ostfront, Sommer 1943: Ein General bei seinen Truppen. Neben der geöffneten Tür sieht man deutlich den Fahrersitz-Lehnenrahmen. Im Hintergrund unter den Bäumen, vor Fliegersicht getarnt, die schweren Halbketten-Zugmaschinen 18 t Famo (Sd. Kfz. 9).

Ostfront, Südabschnitt, Sommer 1943: Der Feldgendarmerie-Trupp einer Division bei der Rast. Der vordere Kastenteil ist um das Ersatzrad mit Wegweisern bestückt. Auf dem rechten vorderen Kotflügel ein 20-Liter-Einheitskanister in einem Haltegestell Marke »Eigenbau«. Auch die Radkappe des Ersatzrades ist eine eigene Erfindung. Rechts im Hintergrund ein Kübelwagen Daimler-Benz 170 V.

Süd-Frankreich, Mittelmeerküste, Sommer 1943: Ein General der französischen Vichy-Regierung nimmt von einem VW-82/trop aus die Parade seiner Truppen ab.

7

Der harte Winter von 1941/42, der die Truppen an der Ostfront unvorbereitet trifft, verstärkt bei dem OKW den Wunsch nach einem schneegängigen Fahrzeug. Und was liegt näher, als den Allround-VW-Kübelwagen zum Motorschlitten umzubauen. *Ingenieur Herbert Kaes*, Professor Porsches prominenter Versuchsfahrer ist auch dabei. Sein Bericht schildert die einzelnen Etappen auf der Suche nach einem Fahrzeug auf der Basis des VW 82 und VW 166, das den Ostfront-Truppen Erleichterung bei den schweren Winterkämpfen bringen könnte:

Betr. Motorschlitten Typ 82 u. 166
Nach Besprechung mit Oberst Rothardt sowie Dipl.-Ing. Knacke/HWA wurden in der Zeit vom 22. 6.–24. 6. 42 im Beisein von Oberinspektor Stöckmann/HWA und Major Rieger/Vers. Kraft Kummersdorf Schneefahrversuche am Großglockner unternommen.
Die grundsätzliche Frage des HWA/Oberst Rothardt lautete, ob ein KdF-Wagen auf Schneekufen gestellt oder durch günstige Formgebung des Aufbauunterteiles angetrieben mit an den Laufrädern montierten Flossen im tiefen Schnee fortbewegt werden kann.
Da Typ 166 im Unterteil der Wanne vollkommen glatt ist, wurde lt. Rücksprache mit dem HWA nach Änderung des vorderen Abdeckbleches in Kufenform mit leichter Abschrägung ähnlich einem Boot zum Zerteilen des Schnees Wagen 166/3 verwendet. Als Antriebsflossen wurden Greiferräder nach SK 6388 und zum Vergleich Greiferräder, hergestellt von der Fa. Rieger & Dietz, verwendet.

Versuchsergebnis:
Im Versuchsgelände am Großglockner (Hochtor) liegt ein etwa 1 m tiefer nasser 550 kg/m^3 schwerer Schnee. Beim Einfahrversuch mit Grobstollenrädern 200/12 Dunlop graben sich die Räder sofort ein. Fahrzeug muß auf Straße geschleppt werden.
Bei den Versuchen brach die linke Doppelgelenkwelle Z. Nr. 87.43.52 an der Faust ab.

Typ 166 mit Reifen 5.25/16 und Greiferrädern
Mit den Greiferrädern beider Ausführungen läuft das Fahrzeug auf den Flossen noch in der Ebene; bei kleinstem Widerstand (Schneehaufen) oder Steigung 5–8 % graben sich die Greifer ein und Fahrzeug sitzt an Wanne auf und muß herausgezogen werden. Damit die Greifer nicht so weit einsinken, wurde die Bodenfreiheit durch Entspannen der Federung und Abbau der Reifen verkleinert. Fahrzeug ist auch in diesem Zustand nicht zu bewegen. Die von der Fa. Rieger & Dietz hergestellten Greiferräder sind in ihrer Wirkung durch günstigere Form besser.

Beobachtung:
Trotz des Geländeganges drehen die Greiferräder viel zu schnell. Bei niederen Drehzahlen tritt ein zu großer Rollwiderstand auf und Motorgrenze wird erreicht. Angefahren werden muß mit großer Motordrehzahl, wodurch die Greifer sich sofort einschaufeln und Fahrzeug aufsitzt. Ein Fahren in diesem Zustand ist nicht möglich. Die vom Waffenamt gestellte Frage, ob ein Motorschlitten mit Greiferrädern fahrbar ist, wurde noch durch folgende Zugkraftmessungen verneint:

Rollwiderstand mit Greiferrädern auf Schnee stehend	150 kg
Rollwiderstand mit Greiferrädern auf Wanne aufsitzend	550 kg
Zugkraft des Fahrzeuges ohne Räder und Greifer auf Wanne aufsitzend beim Anzug	800 kg
Zugkraft des Fahrzeuges ohne Räder und Greifer auf Wanne aufsitzend in Bewegung	400 kg
Einsinktiefe der Fahrzeugwanne (Fahrzeug–900 kg)	8 cm

Schlußfolgerung
Da bei einem für den vorliegenden Versuch günstigen Schnee schon 150 kg Rollwiderstand genügen, um den Schnee wegzuschaufeln, ist es unmöglich, einen Motorschlitten mit einer Zugkraft von im Stillstand 800 kg, in Bewegung 400 kg durch Greiferräder mit einigermaßen günstigem Wirkungsgrad vorwärts zu bewegen. Ein Spezialfahrzeug mit überdimensionierten Greifern und entsprechend großer Untersetzung würde einigermaßen Erfolg versprechen, doch niemals die Leistung eines Raupenfahrzeuges erreichen.

Allgemeines über Fahren von Rad- und Raupenfahrzeugen in tiefem Schnee
Radfahrzeuge haben nie die Fahreigenschaft in tiefem Schnee wie ein Raupenfahrzeug, doch kann durch:
1. starke Untersetzung (bei voller Motordrehzahl nicht über 3 km/h. Fahrgeschwindigkeit, Planetenantrieb usw.)
2. genügend große Räder (wobei zu unterscheiden ist, ob Fahrzeug als Zugmaschine oder zum Tragen von Lasten Verwendung finden soll. Als Zugmaschine muß Fahrzeug große Bodenfreiheit mit spez. Auflagedruck von 0,3 und als Lastfahrzeug nicht über 0,15–0,2 haben. Um den genügend kleinen Auflagedruck zu erzielen, ist zur besseren Fahreigenschaft ein Vergrößern des Raddurchmessers vorteilhafter als ein Verbreitern der Laufräder).
3. günstige Formgestaltung der Aufbauvorderteile, um ein Anstauen von Schnee zu verhindern (durch Anbau von Kufen und Schneepflügen)
bessere Leistung erzielt werden.

Versuche des HWA am Großglockner:

Schneekufen:
Um bessere Fahreigenschaften im tiefen Schnee mit den derzeitigen Raupenzugmaschinen zu erreichen, wurden statt der bisherigen Kufen an den Vorderrädern eine breite Kufe an die Achse montiert. Die Kufenbreite reicht jeweils bis 10 cm vor dem Laufrad. Diese Ausführung hat sich der ersten Anordnung weit überlegen gezeigt. Die Kufe muß aber federnd befestigt sein. Bei der Zugmaschine an querliegender Vorderachsfeder montiert.

Schneepflug:
Kombiniert mit der Schneekufe wurde ein Pflug

entwickelt, der bei auftretendem Pulverschnee denselben beiseite schiebt und ein Aufsitzen verhindert.« Zuffenhausen, 16. 11.42

An das Heereswaffenamt
Wa Prüf 6 / I c
Berlin W 35 Tirpitzufer 72–76

Betrifft: Zugkraftmessung am KdF-Wagen im Schnee
Bei der Entwicklung des Typ 155 (Wagen Typ 82 mit Hinterachsantrieb durch Schneeraupe) wurden Untersuchungen angestellt, ob die Vortriebsleistung des Wagens durch Aufsetzen der Vorderräder auf Schneekufen beeinträchtigt wird.
Bei den Versuchen stellte sich heraus, daß auf einer durch Pferdeschlitten und Fußgänger festgetretenen Fahrbahn, auf der ein Fußgänger kaum einsinkt, an den Vorderrädern angebrachte Schneekufen, etwa 30–50 % größere Zugkraft erfordern als bei demselben Wagen ohne Schneekufen. Es ist bei dieser Schneeart und bei dieser Ausbildung der Schneekufen die gleitende Reibung der Schneekufen wesentlich größer als die Rollreibung der Gummiräder. Ist jedoch der Schnee so locker, daß die Vorderräder tiefer als etwa 10 cm einsinken, so verschieben sich die Verhältnisse zugunsten der Schneekufen. Dies bestätigten Versuche mit dem Typ 155, wo bei aufgesetzten Schneekufen in lockerem Schnee das Fahrzeug durch den Raupentrieb noch ohne weiteres fortbewegt werden konnte, während es ohne Verwendung von Schneekufen vorne so tief einsank, daß die Raupe infolge der notwendigen größeren Vortriebsleistung durchrutschte und sich eingrub. Heil Hitler!

Dr.-Ing. h. c. F. Porsche KG

Bericht
über Schneefahrversuche mit der Schneeraupe III/2 auf der Zuflucht/Schwarzwald am 20. Januar 1943.

Schneebeschaffenheit: 0,6 bis 1 m tiefer gesetzter, nasser Schnee.

Beobachtungen:
Übersetzung: Die Wagengeschwindigkeit beträgt bei einer Motordrehzahl von 3000 U/Min. im 4. Gang 30 km/h. Diese Geschwindigkeit ist für Schneefahrt weit ausreichend und darf wegen Überbeanspruchung der Kette sowie des Antriebs nicht überschritten werden. Bei angegebener Schneebeschaffenheit kann in der Ebene noch knapp mit dem 4. Gang gefahren werden (mit Schneekufen). Da sich der Motor besonders in den niederen Gängen leicht überdrehen läßt, wäre eine Geschwindigkeitsbegrenzung durch Anbau eines Drehzahl- oder Unterdrucksreglers vorzuschlagen. Zumindestens sind die Höchstgeschwindigkeiten am Tachokopf aufzuzeichnen.

Laufwerk: Die Fahreigenschaft kann man als gut bezeichnen, doch müßte jetzt unbedingt festgelegt werden, ob die Raupe nur für Schneefahrt oder, wie es beim Fronteinsatz praktisch sein wird, zeitweilig für Fahrten auf versumpften oder gepflasterten Straßen usw. Verwendung finden soll. Für den letzteren Fall wäre für dieses Fahrzeug eine Protzhaken-Kupplung vorzuschlagen, an der der Wehrmachts-Einheitsanhänger befestigt werden könnte. Auf diesem würden Raupenantrieb sowie Schneekufen für Vorderräder bei Radfahrt Platz finden. Bei einem Auffahrversuch auf der 28 % steilen Straße von Oppenau zur Zuflucht bei 10–15 cm tiefem nassen Schnee

ging die Raupe sofort durch und wurde von einem Schwimmwagen Typ 128 mit Reifen 200/12 und an den Vorderrädern Schneeketten einwandfrei hinaufgezogen. Auf der ungeräumten Straße von Zuflucht nach Schliffkopf und Ruhestein mit 1 bis 1,3 m tiefem Schnee war mit Raupenfahrzeug gut zu fahren. Das Radfahrzeug Typ 128 blieb sofort stecken.
Für Schneefahrt sind zusätzliche Greifer in Messerform oder mindestens Ausbildung der Glieder mit Stollen zur vollständigen Ausnutzung der Kette erforderlich.
Wie die Kontrolle zeigt, sind Teile wie Raupenachse mit Laufrollenlager (im Laufe der Erprobung einige Male gebrochen), Lager (nach 900 km Laufzeit starke Abnützungserscheinungen), den erforderlichen Beanspruchungen noch nicht gewachsen. Die Abnützung der Laufrollenlager ist z. T. noch durch die ungefederte Aufhängung bedingt. Beim schnelleren Fahren oder bei Geländeunebenheiten schlägt die Kette auch bei straffer Spannung an den Verbindungslaschen zur Raupenachse sowie seitlich an den Stützlaschen an. Abgesehen von dem dadurch hervorgerufenen Krach, wird die Kette sowie die Raupenachse selbst, da ja jeweils ein Kettenglied kurz hängenbleibt, übermäßig beansprucht. Die auftretenden Brüche und auf der Zuflucht Verbiegen der rechten Raupenachse sind bestimmt durch das Anschlagen beschleunigt worden. Die Verbindungslaschen müssen verlängert werden, oder es ist zu versuchen, durch Anbringung von Abweisblechen das Anschlagen der Kette zu verhindern. Die Kette darf auch allgemein nicht mehr so sehr gespannt werden.
Die Abnützung der Gliedergelenkbolzenbohrungen und Gelenkbolzen von 0,1 bis 0,3 mm nach rund 900 km Laufzeit ist nicht schlecht, sollte aber zur größeren Sicherheit durch stärkere Ausführung verbessert werden.
Die nachträglich angebauten Schneeabweiser am Treibrad haben sich bestens bewährt. Das Durchbohren der Treibräder an der Lauffläche blieb ohne Erfolg.

Einzelraupen-Handbremse: Durch die Abbremsung einer Raupe wird ein einwandfreies Lenken im Schnee erreicht. Die Wirkung hat nicht nachgelassen. Schäden traten keine auf.

Schneekufen: Beim Fahren im tiefen Schnee ist der Anbau von Kufen an die Vorderachse beim jetzigen Vorderachsdruck unbedingt erforderlich. Die Kufen, auf denen die Vorderräder stehen und die man mit Ketten befestigt hat, sind verwendbar, aber viel zu schwer. Auch ist ein Lenken bei gefrorenem Boden kaum möglich.
Vorgeschlagen wird die pendelnde Aufhängung der Kufen ähnlich der Ausführung an Zugmaschinen, wobei die Räder durch die Kufen am Boden laufen. Eine andere Lösung wäre eine breite Kufe in der Mitte der Vorderachse, wie sie neuerdings versuchsweise an Zugmaschinen erfolgreich erprobt wurde. Nachteilig bei letzterer Ausführung ist, daß bei der Zugmaschine mit Starr-Achse die Vorderachse federt, während beim Volkswagen die Achsrohre starr sind.

Schaltkupplung: Beim Umschalten während der Schneefahrt riß die Schaltkupplung im Gummi aus. Beim Fahren im Schnee tritt ein sehr großer Rollwiderstand auf, der ein schnelles Zuschalten der einzelnen Gänge erforderlich macht. Hierbei wird die Schaltkupplung überbeansprucht. Für das Raupenfahrzeug wäre der Einbau einer Stahlblechkupplung vorzuschlagen.

Motorunterschutz: Die Gebläse-Abluftverblechung wird bei Schneefahrt sofort eingedrückt. Da die neuen Typ 82 mit Wagenheizung ausgerüstet werden, muß die Verblechung dementsprechend geschützt sein.

Kotflügel: Die verbreiterten hinteren Kotflügel bewähren sich gut. Zu überlegen ist, ob für die Serie die Schneekufen bei Nichtverwendung auf diesen verstaut werden könnten.

Der knochentiefe Morast auf der Rollbahn, Unbilden der Natur, das feindliche Feuer vom Boden oder aus der Luft und vor allem der schlimmste Feind eines jeden Kraftfahrers, das Verkehrschaos beim Rückzug, machen auch dem *Obergefreiten Karl Eisner* zu schaffen, der mit seiner Einheit vor der Roten Armee zurückweicht:

»Inzwischen war der Befehl gekommen, die Front hinter den Slutsch beiderseits Staro-Konstantinoff zurückzunehmen. Die Rollbahn war durch Kraftwagen und Bespannfahrzeuge derartig verstopft, daß man in drei Stunden nur rund einen Kilometer vorankam; denn im Ort mußte der gesamte Verkehr über eine einzige Brücke geschleust werden.

Bis zum Abend kamen wir mit unserem VW-Kübel auf wenige Kilometer an die Brücke heran. Hier war der Stau so groß, daß wir wohl nicht vor dem folgenden Tage über den Fluß gelangen würden. Zu allem Übel standen unsere Fahrzeuge in diesem Durcheinander noch getrennt, zwischen fremden Wagen und Geschützen eingekeilt. Es gelang uns aber, einen Teil der Fahrzeuge in eine Kolonne zu bringen, die der Spieß anführte, und aus dem Rest zwei weitere Kolonnen zu bilden, von denen die eine der Wachtmeister Müller, die andere ich übernahm.

Vor der Brücke warteten die Fahrzeuge in 10 bis 12 Reihen nebeneinander. Die beiden anderen Kolonnen hatten Glück und kamen weiter, während ich mit meinem Kübel weder vorwärts noch rückwärts konnte. Na, ich ergab mich in mein Schicksal, denn einmal mußte es ja Luft geben. An Schlaf dachte bei der Kälte natürlich niemand. Gegen zwei Uhr kam der Spieß zu Fuß zurück und sagte mir, ich müßte versuchen, möglichst schnell die Brücke zu erreichen, denn um 6 Uhr würde man sie sprengen, und alles was dann nicht auf dem anderen Flußufer sei, müsse liegen bleiben und auch vernichtet werden.

Die Parole von der geplanten Brückensprengung – erst später stellte es sich heraus, daß es nur eine Parole war – drang auch zu den anderen Fahrzeugen, und jeder setzte alles daran, um rüber zu kommen. Ich ging nun erst einmal die fehlenden 1000 m zur Brücke vor, um zu sehen, wie es dort aussah. Einem meiner Unteroffiziere gab ich den Befehl, die Kolonne zusammenzuhalten, und sowie es weiterging, sie dicht aufgeschlossen vorzuziehen. An der Brücke würde ich schon Rat wissen.

Vor der Brücke war es panikartig, die Fahrer der Panjewagen schlugen sich gegenseitig mit der Peitsche, um den Übergang zuerst zu erzwingen, mit einem Wort, es war ein tolles Durcheinander: Kein Offizier und kein Organ der Verkehrsregelung, jeder wurschtelte für sich allein. An der Brücke fand ich dann noch einige Feldwebel von anderen Einheiten, und wir einigten uns, die Sache selbst in die Hand zu nehmen, jeder natürlich mit dem Gedanken, daß seine Kolonne zuerst dabei war. Nachdem wir uns verständigt hatten, in welchen der zehn Reihen nun unsere Kolonnen standen, und in welcher Reihenfolge wir sie ablaufen lassen wollten, versuchten wir mit viel Geschrei wieder Ordnung und Disziplin herzustellen. Es war beinahe aussichtslos, und erst als wir die

Pistolen zogen, konnten wir uns wieder durchsetzen. Mit der schußbereiten Waffe in der Hand ging nun der Ablauf einigermaßen reibungslos.
Vor der Brücke war alles so verschlammt, daß mehrere Fahrzeuge steckenblieben. Um nicht alles den Russen preiszugeben, spannten wir kurzerhand die Pferde aus und kippten die Fuhrwerke samt der Ladung ins Wasser. Anders ging es in dieser Lage nicht. Mit dauernden Blicken auf die Uhr und dem Gedanken an die Brückensprengung ließen wir die Reihen ablaufen, bis auch endlich meine Wagen kurz nach 4 Uhr glücklich auf dem anderen Ufer waren. Nun konnte ich mich nicht mehr um die Brücke kümmern, denn ich mußte meine Kolonne wieder dem großen Haufen zuführen. Diese Nacht vom 8. zum 9. März 1943 wird mir immer im Gedächtnis bleiben. Es war nur gut, daß sich diese Stauungen in der Nacht auswirkten, denn wenn der Russe das am Tage gemerkt und mit seinen Schlachtfliegern dazwischen gehalten hätte, dann wäre etwas los gewesen! Man durfte gar nicht daran denken! Nach längerem Suchen fand ich schließlich unseren Haufen wieder und konnte meine Kolonne wohlbehalten beim Spieß zurückmelden. Jetzt waren wir also auf der großen Rollbahn von Staro-Konstantinoff in Richtung auf Proskuroff.«
In den ersten Julitagen des Jahres 1943 findet bei der Dr. Porsche KG in Stuttgart-Zuffenhausen eine für die Weiterentwicklung der Varianten des VW-Kübelwagens äußerst wichtige Besprechung statt.

Der Oberregierungsbaurat Dr.-Ing. Hanft vom SS-Führungshauptamt (Amtsgruppe A, Amt 10) präsentiert seine Vorarbeiten zur ›Schienengängigmachung‹ des Typs 82, die er im Einvernehmen mit den Firmen Draisinenbau Dr. Alpers & Co., Hamburg und Getefo, Berlin schaffte. Nach eingehender Diskussion, die zwei Tage dauert, an der auch Porsche jun. teilnimmt und zahlreiche technische Fragen erörtert werden, verabschiedet man sich mit folgender Feststellung:
Porsche erklärt sich bereit, bis 21. 7. 1943 einen detailreifen Entwurf zur Fertigungsbesprechung vorzulegen. Wenn irgendwie möglich, wird Porsche 2 Getriebe für ein Fahrzeug für Versuchszwecke selbst anfertigen. Um die Dringlichkeit der Durchführung nicht hinter den bei Porsche in großer Anzahl vorliegenden DE-Aufträgen zurückstehen zu lassen, wird vorstehend erteilten Aufgaben von seiten des SS-Führungshauptamtes, Amtsgruppe A, Amt 10, die Dringlichkeitsstufe DE sowohl für die Konstruktion als auch für die Anfertigung von 2 Probegetrieben erteilt.
Das SS-Führungshauptamt hat für die Erprobung der Schienenräder den Standort Hamburg unter Fühlungnahme mit Draisinenbau und de dortigen Reichsbahndirektion vorgesehen; die Erprobung soll in etwa 10 Tagen stattfinden. Um jedoch auch das Umkehrgetriebe nach Fertigung im engsten Einvernehmen mit Porsche zu erpr ben, erscheint es notwendig, die Getriebeversuche im Standort Stuttgart zusätzlich durchzuführen. Porsche setzt sich wegen der Probestrecke mit der Reichsbahndirektion Stuttgart in Verbindung, um derartige Versuchsfahrten durchführen zu können. Zweckmäßigerweise wird für die Stuttgarter Versuche der Versuchswagen benützt, der in Hamburg läuft. Zwei Monate später, am 23. September 1943, wird das Schienengefährt einigen Offiziellen, darunter Generalmajor Philipps vom HWA und Oberst Thomale, in Vertretung von Generaloberst Guderian, vor-

geführt. Die Herren sind begeistert, die Nützlichkeit dieses Fahrzeugs für die kämpfenden Truppen steht außer Zweifel, jedoch die bürokratischen Mühlen mahlen langsam. Da greift Dr.-Ing. Hanft selbst zur Feder und schreibt an die Feldpostnummer 12 000, eine Sammeladresse, über die jedermann der Kriegsführung nützliche Ideen, Anregungen oder Erfindungen vorschlagen kann:

Stuttgart, den 5. 10. 1943

Ob.Reg.Baurat Dr.-Ing. Hanft
SS-Führungshauptamt (Amt X)

An
Feldpostnummer 12 000.

Betreff: Vorschlag zur Schienengängigmachung von Kraftfahrzeugen

Nachstehend wird der Vorschlag gemacht, den Volkswagen mit einfachsten feldbrauchbaren Mitteln schienengängig zu machen. Durch Spurkranzscheiben, ähnlich Stahlblechrädern, werden die Wagenräder durch die Spurkränze auf der Innenseite der Schiene geführt. An Rädern und Luftdruck wird nichts geändert. Der Unterschied zwischen Wagenspur- und Schienenspurweite wird durch Umdrehen der Räder und die Gestaltung der Spurkranzscheiben erreicht. Dadurch vergrößert sich die Spurweite um 2 × 86 mm = 172 mm.
Lenkung und Schwingachse werden bewußt nicht festgestellt. Das Fahrzeug wird nicht wie ein reines Eisenbahnfahrzeug von der Schiene aus gelenkt, sondern durch leichten Lenkradanschlag kann ein Reiben und etwaiges Quietschen infolge Anliegen des Spurkranzes am Schienenkopf vermieden werden. Dadurch läuft dann der Luftreifen sehr leicht auf der Schiene.
Bei dieser Lösung der Schienengängigkeit ist die allgemein erhobene Forderung erfüllt:

1. keinerlei Änderung der normalen Serienfertigung des Wehrmachts-Volkswagens.
   Ferner zeigt die Lösung folgende weitere Merkmale, welche ihre besondere Feldbrauchbarkeit begründen:
2. einfachste und schnellste Fertigung durch Pressen von vier Spurkranzscheiben je Kfz.,
3. rasche Montage der Verschienungseinrichtung in 8–12 Minuten. Der Wagen ist mit dem Normalwagenheber aufzubocken, die Räder einer Wagenseite sind zu gleicher Zeit abzumontieren, die Spurkränze aufzusetzen, die Räder wieder aufzustecken und mit etwas verlängerten Radmuttern wieder festzuschrauben,
4. vollkommene Fahrtsicherheit bis zur Höchstgeschwindigkeit des Volkswagens 80 km/h. Die Nachprüfung des eisenbahntechnisch einwandfreien Laufens der Reifen und Spurkränze auf den Schienen erfolgte durch die Deutsche Reichsbahn,
5. leichte Umkehrmöglichkeit an einem schienengleichen Übergang in 1 Minute durch Herausfahren aus den Schienen mit und ohne Schienenschuh und Beschreiben einer Schleife bei aufgesetzten Spurkränzen,
6. sehr gute Bremsfähigkeit sowohl auf nasser als auf trockener Schiene. Die Bremsverzögerungswerte sind gleich und besser als auf Asphalt. Bei der Bremsung kommt die sehr gute Führung der Spurkränze zur Wirkung und macht ein seitliches Wegrutschen wie auf Asphalt unmöglich,
7. geringes Gewicht der Spurkranzscheiben, je Scheibe etwa 15 kg, gesamt etwa 60 kg,

8. gute Unterbringungsmöglichkeit der Spurkranzscheiben unter dem Reservereifen, ohne den Gepäck- oder Mannschaftsraum zu beeinträchtigen.

Diese einfache Art der Verschienung von Kraftfahrzeugen ist bei Spurweiten kleiner als 1435 mm anwendbar.

Anläßlich einer durch WA Prüf 6 veranlaßten Vorführung am 23. 9. 43 auf dem Eisenbahnpionierübungsplatz Klausdorf fand diese Lösung ihrer Einfachheit halber ganz besondere Anerkennung.

Die Vorführung erfolgte mit Versuchsspurkranzrädern, die aus dem vollen gedreht waren. In der Zwischenzeit sind die Vorarbeiten für das Pressen der Scheiben aus 5 mm bzw. 3,5 mm Stahlblech so weit gediehen, daß der Serienanfertigung nichts mehr im Wege steht. Ich bitte gehorsamst um Prüfung dieses von mir entwickelten Gedankens der Schienengängigkeit von Kraftfahrzeugen.

gez. Dr.-Ing. Hanft.

Die Hoffnung des Oberregierungsbaurates Dr. Hanft, daß jene dem OKW unterstellte Prüfungskommission, die sich unter der Feldpostnummer 12 000 verbirgt, unbürokratisch und ohne den üblichen Amtsweg seine Idee zu realisieren verhilft, hat sich erfüllt. Im Frühjahr 1944 läuft der Kübelwagen 82, schienengängig, (VW schg) bei der Truppe. Leider viel zu wenig und zu spät.

Am Montag, dem 4. September 1944, weilt Professor Porsche im Führer-Hauptquartier Wolfsschanze in den Wäldern um das ostpreußische Rastenburg. Just einen Tag zuvor hat die 2. britische Armee Brüssel eingenommen, und die Sowjets stehen an der Weichsel vor Warschau.

Professor Porsche erscheint im Hauptquartier mit Ingenieur Herbert Kaes und den Monteuren Rupprecht und Geisinger wegen des Kübelwagens, – genauer gesagt – VW Typ 276, der auf der Basis des VW 82 mit Protzen für Pak 37 und Munitionsanhänger konstruiert wurde. Zur Vorführung kommen höchstpersönlich Adolf Hitler samt Generalfeldmarschall Keitel, Panzer-Spez. Generaloberst Guderian und SS-Hauptsturmführer Skorzeny. Dies läßt erkennen, welche Gewichtigkeit man dem unscheinbaren Kübelwagen beimißt. Das Fahrzeug wird von *Herbert Kaes* gesteuert, und in seiner Aktennotiz ist darüber zu lesen:

Vorführung des Typ 276 mit Pak 37 und Anhänger nach Zeichnung Nr. 276.00.010 im Hauptquartier am Montag, dem 4. Sept. 1944.

1. Bei der Vorführung vor dem Führer waren anwesend die Herren:
   Generalfeldmarschall Keitel,
   Generaloberst Guderian,
   Prof. Dr. Porsche.
   Die Angaben über Änderungen am Fahrzeug sowie über den Anhänger wurden von Herrn Prof. Porsche persönlich angegeben.
   Mit Herrn Oberst Holzhäuer wurde vereinbart, daß eine sofortige Erprobung in Berka bei Eisenach durchgeführt wird. Als Vergleichsfahrzeug wird von der Vers. Kraft. Kummersdorf ein Steyr 1,5 to gestellt.
   Ein Fahrzeug mit Pak und Anhänger kommt vor der Erprobung noch nach Stuttgart, damit, wie mit Herrn Prof. Porsche besprochen, die Anhängerpritsche auf Holzkonstruktion umgebaut werden kann. Die Erprobung soll bereits Ende nächster Woche beginnen. Der bisher für die Erprobung verbrauchte und noch zu benötigende Kraftstoff wird laut Herrn Hauptmann Braun-

Munsinger der Porsche KG überwiesen. Ein Entwicklungsauftrag wird erteilt. Da ein Ausreißen oder weiteres Versinken des hinteren Querträgers durch die angehängte Last bei der Berka-Erprobung möglich ist, wird die Herstellung eines verstärkten Ersatzquerträgers vorgeschlagen.

2. Herr Prof. Porsche hat Sturmbannführer Kempka (Fahrer des Führers) die baldige Lieferung eines Volkswagens mit Holzkohlengeneratoranlage zugesagt. Das Fahrzeug soll in Berlin, Hermann-Göring-Str. 16, abgeliefert werden.
3. Weiter hat Herr Prof. Porsche Herrn Hauptsturmführer Skorzeny die schnellstmögliche Lieferung eines Anhängers zugesagt. Die Maße sollen etwa dem für die Pak 37 gebauten Anhänger entsprechen, eine Nutzlast von 200 kg tragen und mit Reifen 5,25 × 16 ausgerüstet sein. Das Eigengewicht soll so niedrig wie möglich gehalten werden. Evtl. leichtes Verdeck. Verbindung zu Hauptsturmführer Skorzeny über Herrn Dr. Tirolt, VW-Werk.

Stgt.-Zuffenhausen, 7. 9. 1944. Ks/kk.

Programmgemäß werden zwischen dem 19. September und 17. Oktober 1944 auf dem Übungsgelände Berka, in der Nähe von Eisenach, die Versuchsfahrten mit dem VW Typ 276 durchexerziert: abwechselnd mit angehängter Pak 37 oder mit einem Munitionsanhänger. Versuchsweise wird auch ein VW-Kübel 82 mit Hinterachsübersetzung auf seine Eignung als Zugmittel für den Munitionsanhänger und die Pak 37 geprüft.
Die Wagen, vom routinierten Herbert Kaes geführt, kurven insgesamt 738 Kilometer auf der Berka-Rundstrecke Nr. 1 mit einem Durchschnittstempo von 18 km/h, bei nasser Wegstrecke und Kraftstoffverbrauch von 21 Ltr/100 km. Nach Zuffenhausen zurückgekehrt, verfaßt *Ingenieur Kaes* am 2. November 1944 einen Zwischenbericht über die Erprobung des Typ 276:
»Bei Fahrt auf Landstraße und Reichs-Autobahn hat das Fahrzeug mit angehängter Pak gute Straßenlage. Der Munitionsanhänger führt etwas schlechter und stößt stärker an die Anhängerkupplung. Bei lang anhaltender Bergabfahrt muß zur Schonung der Bremse der niedrigste Gang eingeschaltet und möglichst zwischendurch angehalten werden. Allgemein muß auf nasser Straße besonders bei Bergabfahrt wegen Schleudergefahr vorsichtig gefahren werden.
Mit der im Auftrag des Heeres-Waffen-Amtes versuchsweise eingebauten Serienübersetzung Typ 82 hat das Fahrzeug schlechte Beschleunigung, muß häufig geschaltet werden und läuft auf Landstraße nicht über 65–70 km/h.

Geländeerprobung in Berka. Mit den angebauten Reifen 200 × 16 wird eine verhältnismäßig gute Geländegängigkeit erzielt. Die Getriebeübersetzung ist ausreichend. Der zur gleichen Zeit auf der Rundstrecke zur Erprobung laufende Steyr 1,5 to mit Allradantrieb, einem zulässigen Gesamtgewicht von 4600 kg und angehängtem Geschütz von 1560 kg hat mit Serienbereifung schlechtere (Bodenfreiheit-Reifengrenze), und mit Überreifen gleiche Geländegängigkeit wie Typ 276. Die Steigleistung des Steyr ist auf trockenem Geländeboden (größere Übersetzung) um 1–2° besser. Die Steiggrenze des Typ 276 liegt je nach Bodenbeschaffenheit bis Erreichung der Motorgrenze bei 19°. Bei einer Abfahrt mit angehängtem Munitionsanhänger auf einem 15° nassen steilen Wiesenhang überschlug er sich durch Querstellen und Abrutschen des Anhängers und den hierdurch auf das Fahrzeug wirkenden starken Seitendruck des Wagens. Steile Hänge müssen besonders vorsichtig und unbedingt gerade angefahren werden. Der Steyr mit angehängtem Geschütz ist noch empfindlicher auf Schleudern und Kippen.

Mit versuchsweise eingebauter Hinterachsübersetzung Typ 82 wird eine Steigleistung von 10° erreicht und kann im Gelände (Rundstrekke) nur mit dem 1. und 2. Gang wegen Motorgrenze gefahren werden.

Beanstandungen, Instandsetzung und Neueinbau

Das Fahrzeug ist mit einer *Flatterbremse* ausgerüstet. Nach etwa 3500 km Fahrleistung wurde ein leichtes Flattern der Vorderräder festgestellt, das nach der notwendigen Nachstellung der Vorderachsbüchsen und Lenkung bis jetzt nicht mehr auftrat.

Die linke vordere *Wagenheizungsklappe* blieb auch beim Einstellen der Heizung geschlossen. Die Heizung wurde gerichtet.

Die durch das Überschlagen des Fahrzeugs im Gelände aufgetretenen Schäden wie verbogenes linkes Achsrohr, Hinterachsstrebe, Manschetten, 1 Felge und Lenkrad wurden erneuert, der verbogene Hinterachsanschlag und beschädigte Kotflügel gerichtet.

Die Funktion des *Drehzahlbegrenzers* ist einwandfrei. Lediglich die Zug- und Mitnehmerhebel am Vergaser blieben bei km 4100 hängen und mußten nachgerichtet werden. Eine Verbesserung dieser Hebel mit Lagerung ist erforderlich. Die Kugel des Reglerhebels wurde jeweils nach 500 km geschmiert.

Nach etwa 3500 km Fahrleistung rutschte die *Motorkupplung* trotz richtiger Pedaleinstellung bei starker Belastung durch. Bei Kontrolle wurde festgestellt, daß die Schrauben der Kupplungsdruckplatte locker waren (Montagefehler) und dadurch die Pressung zur Übertragung des erforderlichen Drehmoments der sonst einwandfreien Kupplungsscheibe fehlte. Sicherheitshalber wurde eine neue komplette Kupplung eingebaut.

Ab Fahrzeug-km 3700 davon 500 km Gelände wurde ein Versagen und starkes Krachen des ZF-Sperrdifferentials festgestellt. Das Sperrdifferential wurde bei 4100 km ausgebaut. Die Führungsbahn der Gleitsteine an der Kurvenscheibe ist sehr abgenützt, und die Laufbüchsen haben sich festgefressen. Ein neues ZF-Sperrdifferential wurde eingebaut und das alte an die Herstellerfirma zur Untersuchung gesandt. Die *Bremsen* mit Seilen wurden nach Werkserprobungsfahrt bei km 1407 neu eingestellt. Während der Geländeerprobung in Berka mußten die Bremsen dreimal nachgestellt und km 4150 wegen zu starker Abnützung an den Hinterrädern neu belegt werden. Die Wirkung der Bremse ist in gutem Zustand und bei normaler Beanspruchung ausreichend. Mit der angehängten Last muß selbstverständlich zur Schonung der Beläge mit den unteren Getriebegängen mitgebremst werden.

Durch das Fehlen der früher serienmäßig an den vorderen Kotflügeln angebrachten *Schmutzabweiser* wird das Fahrzeug selbst wie auch der Anhänger sehr stark verschmutzt. Es wurden Schmutzabweiser aus Blech angebaut.

Anhängerkupplung mit Halterung. Von der Firma Ade, Waltershausen, wurde eine vereinfachte Anhängerkupplung angebaut und der Haltebock verstärkt. Diese Änderung hat bisher etwa 500 km im Gelände mit angehängter Last gehalten. Der Kupplungshaltebolzen ist im

Haltebock etwas ausgeschlagen, die Anhängerkupplung selbst in Ordnung. Der Querträger am Fahrzeug ist durch Last etwa 7 mm durchgebogen. Nach 650 Gelände-km ist der Querträger an den 4 Befestigungsschrauben des Haltebockes für Anhängekupplung ausgerissen und wurde durch eine 5 mm starke Blechplatte verstärkt und elektr. verschweißt.

Munitionsanhänger. Bei Geländefahrt wird der weiche Lehmboden zwischen Rad und Kotflügel gepreßt, wodurch die Räder feststehen. Es wurden beide Kotflügel abgebaut. Um das Hochschleudern der nassen Erde und ein dadurch auftretendes Verschmutzen des Fahrzeuges und Anhängers zu vermeiden, wird der Anbau kleiner, vom Rad möglichst weit entfernter Kotflügel vorgeschlagen.
Wie schon erwähnt, hat die Pak 37 gegenüber dem Anhänger bessere Straßenlage und Geländegängigkeit. Die Vorteile sind durch folgende bauliche Unterschiede gekennzeichnet: Größere Räder, dadurch bessere Bodenfreiheit, lange und gerade, in angehängtem Zustand waagerechte Deichsel und ein besser ausgeglichener Deichseldruck. Durch die schräge und kurz gehaltene Deichsel am Anhänger wird die Anhängerkupplung bei Schubkraft hochgedrückt und schlägt stark. Um den Einschlagwinkel des Anhängers zu vergrößern wurden die Haltegriffe der Deichsel um je 25 mm gekürzt.«
Die Versuchsfahrten sind zu dieser Zeit nicht ohne Risiko. Als Herbert Kaes Ende August 1944 im saloppen Zivil mit dem VW 276 und angehängter Pak 37 in Richtung Katschberg unterwegs ist, wird er in Radstadt von einem eifrigen General namens Ringl aus dem Wagen geholt und soll auf dessen Geheiß arretiert werden, da er »als Zivilist ein Wehrmachtsfahrzeug mit Geschütz gefahren hat«.
Im Herbst 1944 muß Porsche auf sanften Druck des Reichsrüstungsministers Albert Speer sein Stuttgarter Konstruktionsbüro wegen der Gefahr von Bombenangriffen in das abgelegene Gmünd nach Kärnten/Österreich verlegen. Auch Teile des VW-Maschinenparks aus Wolfsburg werden ins Erzgrubenbergwerk nach Longuich, nahe der belgisch-luxemburgischen Grenze evakuiert. Zunächst sind es zweihundert wertvollste Werkzeugmaschinen, die mit der Reichsbahn an ihre neuen Standorte geschafft werden. In Grubenstollen installiert, fallen sie kurz danach den vorrückenden amerikanischen Truppen in die Hände. Übrigens im Oktober 1944 – dank diverser organisatorischer Maßnahmen Speers – steigt die Herstellung der VW-Kübel trotz der immer stärkeren Bombenangriffe auf 1380 Wagen und im Januar 1945 sogar auf 2030. Als man jedoch im darauffolgenden Monat jegliche Lieferung von Rohmaterialien wegen der kritischen Lage an den Fronten stoppt, erreicht die Produktion ihren absoluten Tiefststand.
Während die amerikanische Panzerspitze sich Anfang April 1945 Wolfsburg nähert, baut man geschwind die wichtigsten Teile der noch im VW-Werk verbliebenen Maschinen aus, verstaut sie in Kisten und verbirgt sie in den unterirdischen Kabelschächten, eine – wie sich später herausstellt – für die Existenz des Werkes äußerst wichtige Maßnahme.
Am Abend des 10. April, so gegen 20.45 Uhr, werden die Bewohner der »KdF-Stadt« durch Panzeralarm aufgeschreckt. Die 9. US-Armee (Gen.Lt. Simpson) steht kurz vor dem benachbarten Fallersleben. Dies ist ein Signal für mehrere Tausend zwangsrekrutierte Fremdarbeiter aus Ost und West, Wolfsburg und das herrenlose Volkswagen-Werk unsicher zu ma-

chen, aber die US-Panzer blieben aus. Die Amerikaner können Wolfsburg nicht erobern, da die »KdF-Stadt« auf den Karten, die sie besitzen, gar nicht eingezeichnet ist.
Um das Schlimmste zu verhüten, eilt der Chefinspektor des VW-Werkes, Brörmann, gegen Mitternacht nach Fallersleben und bittet die Amerikaner, Wolfsburg zu besetzen. Kurz darauf rollen einige Sherman-Panzer in Richtung Stadt und postieren sich vor dem Werkstor.
Damit geht ein Kapitel in der Geschichte des Volkswagens zu Ende.

Süd-Italien, Sommer 1943: Ein Soldat flickt den Schlauch des Hinterrades. Darauf liegt die Luftpumpe, daneben eine Werkzeugtasche, ganz rechts unter dem Kotflügel die Hinternabe und die Bremstrommel.

Mittelmeerraum, Sommer 1943: Ein General auf Inspektionsreise mit einem VW-82. Der Wagen hat Kronprinz-Sandreifen.

Süd-Italien, Sommer 1943: Ein Nachrichten-Trupp beim Einsatz. Rechts dessen Fahrzeug, ein VW-82. Auf dem hinteren, rechten Kotflügel der Spatenverschluß.

Reichsgebiet, Sommer 1943: Der VW-82 einer Panzer-Einheit wird von Hand auf den Eisenbahnwagen verladen. Dabei helfen gut 30 Mann.

Reichsgebiet, 17. August 1943: Auf einem Flugplatz bei Regensburg nach dem amerikanischen Bombenangriff. Vorn ein VW-82 der Luftwaffe.

Ostfront, Sommer 1943: Dieser Schnappschuß bestätigt noch einmal die ungewöhnliche Geländegängigkeit des VW-82.

Italien, Sommer 1943: Nach dem deutschen Rückzug aus der Stadt Scafati. Ein im Stadtzentrum verlassener VW-82.

Süd-Italien, Herbst 1943: Ein neuer VW-82 auf dem Feldflugplatz eines Jagdgeschwaders. Im Hintergrund eine Bf 109 G.

Süd-Italien, Herbst 1943: Blick durch die Windschutzscheibe des VW-82 auf den Feldflugplatz eines Jagdgeschwaders. Links eine BF 109 G.
Auf dem Schaltbrett kann man deutlich erkennen

1. **Scheibenwischer**
2. **Signalknopf**
3. **Sicherungskasten.**
4. **Anlasser-Druckknopf**
5. **Steckdose für Handlampe**
6. **Schalter für Schaltbrettleuchte**
7. **Ladeprüflampe**
8. **Öldruckprüflampe**
9. **Geschwindigkeitsmesser und Kilometerzähler**
10. **Zündschloß**
11. **Anzeigelampe für Winker**
12. **Lichtschalter**
13. **Anzeigelampe für Fernlicht**
14. **Mehrfachschalter**
15. **Sicherungskasten**
16. **Winkerschalter**

Süd-Italien, Herbst 1943: Innenraum eines VW-82 Kübelwagens. Entlang der vorderen Sitzlehnen die Anhaltestange mit Karabinerhaltern. Vorn das Verdeckgestell.

Süd-Italien, Barletta, 1943: Ein Kübelwagen-VW auf dem Marsch mit einer Kradschützen-Einheit.

Süd-Italien, Barletta, September 1943: VW-82 Kübelwagen einer Panzerspäh-Kompanie.

Ostfront, Ukraine, Herbst 1943: Zwischen den Fuhrwerken vor einem kleinen ukrainischen Provinzbahnhof der VW-82.

WL-244 067

Linke Seite oben:
Ostfront, Mittlerer Abschnitt, Winter 1943: Eine Panzer-Einheit auf dem Vormarsch. Vorn ein VW-82 mit teilweiser Wintertarnung.

Linke Seite unten:
Reichsgebiet, Winter 1943: Ein VW-82 der Luftwaffe auf einem Rungenwagen vom Typ »Stuttgart«.

Freising, nördlich von München, Winter 1943: Es geht auch so. Die Soldaten einer Luftwaffen-Einheit heben eigenhändig ihren VW-82 auf den Eisenbahnwagen. ►

Ostfront, Winter 1943/44: Kurz vor dem Aufbruch zum Rückzug. Im Hintergrund ein VW-Kübelwagen. ▼

Ostfront, Winter 1943/44: Auf dem verschneiten Weg steckengeblieben. Die Landser helfen beim Bergen eines VW-82.

Ostfront, Winter 1943/44: Die Nachrichtentruppe der Nachhut auf dem VW-82 vor einer brennenden Ortschaft. Aus den vorderen Fahrzeugen ragen die Antennen der Funkgeräte heraus.

Süd-Italien, Frühjahr 1944: Der VW-82 einer Panzer-Einheit inmitten eines Olivenhains in der Nähe von Anzio. Die umgeklappte Windschutzscheibe und der vordere Kastenteil samt Ersatzrad sind mit einer Schutzplane bedeckt.

Süd-Italien, Frühjahr 1944: Der vordere Kastenteil des VW-Kübelwagens, Typ 82, bekommt mit der Luftdruck-Spritzpistole einen Tarnanstrich. Im hinteren geöffneten Kastenteil befinden sich an der Anhaltestange die vierteiligen Karabinerhalter, darunter auf dem Fußbodenblech der vierteilige Karabinerschuh, daneben der Sammlerkasten. Hinten über dem Motor der Deckel für den Motorraum und Gerätekasten. Zwischen den äußeren rechten und linken Seitenteilen die Hintersitzkissen-Kammer. Die Hintersitz-Lehnrahmen und Kissen sind ausgebaut. Vorn neben dem Preßluftschlauch steht das Vordersitzgestell.

Nord-Italien, Frühjahr 1944: Ein VW-Kübelwagen der Luftwaffe vor der Einfahrt zum Quartier der rückwärtigen Dienststellen.

Der Posten weist dem Fahrer des VW-Kübelwagens den Weg zum Gelände der rückwärtigen Dienststellen.

Italien, Frühjahr 1944: Der Kübelwagen eines Luftwaffen-Stabes vor einem Straßentunnel in den Dolomiten.

Nord-Frankreich, Bretagne, Frühjahr 1944: Ein General mit seinem VW-82 auf der Inspektionsfahrt.

Ostfront, Frühjahr 1944: Ein VW-Kübelwagen der 78. Sturm-Division geht auf die Reise.

Rechte Seite oben:
Italien, Sommer 1944: Probefahrt mit einem gerade instandgesetzten VW-82. Das Fahrzeug hat noch keine Zulassungsnummer.

Rechte Seite unten:
Frankreich, in der Normandie bei Caen, Juni 1944: Ein gegen Fliegersicht getarnter Kübelwagen. Hinten auf dem Verdeck zwei Fallschirmjäger.

◄ Italien, Frühjahr 1944: Ein Kübelwagen im unwegsamen Gelände an der Anzio-Nettuno-Front.

Italien, Frühjahr 1944: Ein Treff zwischen Natur und Technik. Der VW-82 auf einer Dorfstraße am Rande der Apenninen. ▼

Frankreich, Sommer 1944: Ein VW-82 der Luftwaffe auf einer Straße von Grenoble.

Jugoslawien, Sommer 1944: Ein Kübelwagen des Versorgungsdienstes einer Heeres-Einheit auf der Straße von Zadar nach Sebenico.

Frankreich, Normandie, zwischen Alencon und Dreux, Juli 1944: Die Besatzung eines Kübelwagens sucht Schutz vor Tieffliegern.

Frankreich, Normandie, Juli 1944: Noch einmal gut gegangen! Eine Kübelwagen-Besatzung entkam mit knapper Not dem Feuerinferno. Auf der Straße von Alencon nach Dreux, eine von Jabos in Brand geschossene Lkw-Kolonne.

Frankreich, Normandie, Straße von Alencon nach Dreux, Juli 1944: Doch erwischt! Auf dem Weg zu einer Kfz-Instandsetzungswerkstatt haben die Jabos einen Kübelwagen in Brand geschossen.

Norwegen, Finnmark, Sommer 1944: Der Kübelwagen-Park des im hohen Norden an der Eismeerfront eingesetzten Jagdgeschwaders 5. Der VW-82 im Hintergrund links gehört zum III./JG 5 (Eismeer).

Nord-Frankreich, Sommer 1944: Ein Kübelwagen der Luftwaffe am Rande eines Feldflugplatzes.

Frankreich, Sommer 1944: Eine Kraftfahrzeug-Einheit auf dem Marsch.

Ostfront, Südpolen, Sommer 1944: Ein Kübelwagen vor Fliegersicht getarnt, auf einem Waldweg beim Rückzug.

Italien, Apenninen, Sommer 1944: Am Rande des Städtchens Esperia zerstörte Einheiten der 71. Infanterie-Division. Vorn die Trümmer eines VW-82. ►

Ostfront, Baltikum, Sommer 1944: Ein VW-82 einer Heeres-Einheit mit Propaganda-Material.

Holland, September 1944: Der VW-82 einer Luftwaffen-Einheit im Wald in der Nähe von Arnheim.

Holland, Arnheim, September 1944: Drei gefangene Engländer werden zum KG-Lager abtransportiert.

Arnheim, Holland, 19. September 1944: Im Kampf mit britischen Luftlande-Truppen gefallener Fahrer eines VW-Kübelwagens 82 auf dem Utrechtsweg nahe der Oranjestraat.

▲

Holland, September 1944: Ein von den alliierten Jabos zusammengeschossener Verwundeten-Transport. Links ein VW-Kübelwagen mit Rote-Kreuz-Zeichen, rechts ein VW-166 Schwimmwagen.

Mittelmeerraum, Herbst 1944: Ein General mit einem Oberfeldwebel der Feldgendarmerie im VW-82 des Kraftwagen-Werkstattzuges einer Division. Die kurze Fahrtunterbrechung nutzt ein Soldat aus, um an den Wagen ein Durchhalte-Plakat mit der Aufschrift »Trotzdem siegen wir!« zu kleben.

Slowakei, Herbst 1944: Ein VW-Kübelwagen vor der Verpflegungsausgabestelle einer Panzer-Einheit.

Reichsgebiet, im Raum von Köln, Dezember 1944: Ein VW-Kübelwagen in der feldmäßigen Garage.

Ardennen-Offensive, 29. Dezember 1944: Die von den deutschen Truppen wegen Treibstoffmangels zurückgelassenen Fahrzeuge auf einem Feld in der Nähe des Städtchens Foy-Notre-Dame.

Das Ende. Wolfsburg/Fallersleben, Mai 1945: Die durch alliierte Bombenangriffe zerstörten VW-Werkshallen. Rechts die Kübelwagen. ▶

Griechenland, Rhodos, Mai 1945: Nach der Besetzung durch britische Streitkräfte verrichten die Fahrer und Wagen auch weiterhin ihren Dienst. Vorn ein VW-82 des Heeres.

8

DATEN + FAKTEN

Technische Daten leichter Pkw Volkswagen Typ 62/1939

Motor	
Bauart	liegend, gegenläufig
Zylinderzahl	4 (Boxer)
Bohrung	70 mm
Hub	64 mm
Hubraum	984 cm^3
Verdichtung	1 : 5,8
Leistung max	24 PS
Drehzahl max	3200 U/min
Drehzahl normal	3000 U/min
Zylinder Werkstoff	Grauguß
Zylinderkopf Werkstoff	Silumin
Steuerung	hängende Ventile
Anzahl der Ventile je Zylinder	Einlaß = 1, Auslaß = 1
Lage der Nockenwelle	unten
Kurbelwelle	4 Gleitlager
Kolben	Leichtmetall
Zündung	Batterie
Zündfolge	1 - 4 - 3 - 2
Zündkerzen	14 mm, Typ: W 145 T 7 Bosch
Untersetzung der Verteilerwelle	2 : 1
Verteiler	Bosch VE 4 B S 276
Lichtmaschine	Bosch RED 130/6/2600 Al 8p
Anlasser	Bosch RED 0,4/6 L 3 (P)
Vergaser	Fallstrom Solex 26 VFJ
Ventile	Hängend, Stoßstangen und Kipphebel, Zentrale Nockenwelle, Antrieb durch Stirnräder
Kühlung	Luftgebläse
Drehzahl des Gebläses	6000 U/min
Luftmenge	500 l/min
Schmierung	Umlauf
Ölpumpe	Zahnrad
Ölkühler	Röhrenkühler
Kupplung	Einscheiben, trocken, Größe K 10 Fichtel und Sachs
Getriebe	
Gangzahl	vorwärts 4; rückwärts 1
Untersetzung	1. Gang = 1 : 3,60 2. Gang = 1 : 2,07 3. Gang = 1 : 1,25 4. Gang = 1 : 0,80 R. Gang = 1 : 6,6
Antriebsübersetzung	4,43 (7 : 31)
3. und 4. Gang	geräuschlos
Zahnradausführung	1. u. 2. Gg. u. R.Gg. gerade verzahnt Schubschaltung 3. u. 4. Gg.
schräg	schräg verzahnt Stiftschaltung
Wagengeschwindigkeit bei Motor n = 3200 U/min	1. Gang = 18,4 km/h 2. Gang = 32,4 km/h 3. Gang = 54,2 km/h 4. Gang = 83,6 km/h R. Gang = 10,1 km/h
Höchstgeschwindigkeit	83 km/h
Art der Schaltung	Kugel, Mitte Wagen
Fahrgestell	
Rahmen	Drehungssteifer Mittelrohrrahmen mit Gabelung hinten zur Aufnahme des Motors. Hinterachse, Triebling u. Tellerrad
Motorlage	Heckmotor, 3 Punkte gelagert
Triebachsausgleich	Kegelräder, Z. F. oder Rheinmetall selbstsperrend
Schubübertragung	Strebe-Federhebel

Lenkung

Lenkung	Sonderspindellenkung Konstruktion Dr. Porsche
Übersetzung	2,4 Umdr. am Lenkrad = 49° = voller Einschlag der Vorderräder 1 Umdr. am Lenkrad = 20,5° = 42%
Lenkrad	∅ 400 mm
Lenkungslage	links (rechts)
Spurstange, geteilt, oberhalb des Rahmens	
Wendekreis	10 m ∅
Radeinschlag	äußeres Rad = 28°; inneres Rad = 35°

Federung

Vorderradaufhängung	unabhängig gefederte Räder (Dr. P.)
Vorderradfederung	2 Stabfedern
Hinterradaufhängung	Schwingachse mit Strebenabstützung
Hinterradfel erung	1 Federstab, Nonius, einstellbar
Stoßdämpfer	hydraulisch, vorne und hinten

Schmierung

Dauerschmierung, an 7 Schmierstellen Druckschmierung, 4 Vorderachsschwinghebel, 2 Vorderachsbolzen, 1 Lenkung

Bremsen

Bremsen	Wirkungsweise mechanisch durch Seile, Konstruktion Dr. Porsche Durchmesser 230 mm Breite 30 mm
Fußbremse	wirkt auf 4 Räder
Handbremse	wirkt auf 4 Räder
Räder	5 Scheibenräder, je 5 Bolzen
Felgen	(3,25 E x 18)
Reifen	Geländereifen (5,00 - 18)
Aktiver Radius	normal = 345 mm
Spurweite vorne	1356 mm
Spurweite hinten	1356 mm / 1316 mm
Radstand	2400 mm
Kleinster Bodenabstand	unbelastet 260 mm
bel. mit 460 kg	240 mm

Bodenfreiheit

Fußboden Oberkante über	unbel. 280 mm
Boden	bel. 260 mm

Betriebsstoffe

Kraftstoff	Benzin
Kraftstoffverbrauch	9 l/100 km
Kraftstoffbehälter	Fassungsvermögen: 40 l mit Reserve (10 Liter) Lage vorne Zuleitung: Pumpe
Motoröl	Verbrauch 0,12 l/100 km Fassungsverm.: 3 l
Getriebeöl	Fassungsverm.: 2,5 l

Reserverad

Anzahl: 1, Lage vorne im Aufbau

Batterie

Spannung 6 Volt; 75 Amp., Typ DIN Kr 2311 (unter Rücksitz)

Aufbau 3sitzig oder 4sitzig offen
Ganzstahlaufbau
Gerätekasten (für 3sitzigen Aufbau) abnehmbar
Windschutzscheibe umlegbar, Sicherheitsglas
Klappverdeck abnehmbar
Fußboden am Fahrgestell befestigt
Instrumentenbrett: Geschwindigkeitsmesser mit km-Zähler
Öldruckanzeiger
Fernlicht- und Ladekontrollampe
Winker am Vorderteil, Winkerschalter
Scheinwerfer am Vorderteil
2 Einzelscheibenwischer
4 (3) Sitze;

Breite vorne etwa	1060 mm
Breite hinten etwa	1120 mm (560 links)

Gepäckraum hinter den Hintersitzen L:B:H etwa 1150:380:450 mm

Wagenabmessungen:

Größte Länge etwa	3750 mm
Größte Breite etwa	1550 mm
Größte Höhe etwa	1550 mm (mit Verdeck)
Reifengröße	5,00 - 18
Wendekreis ∅	11 m

Wagengewicht: (gerechnet mit Öl)

Fahrgestell mit Motor	etwa 392 kg
Aufbau	etwa 170 kg
Ausrüstung	etwa 80 kg
Gesamtgewicht leer, fahrbereit (mit 40 l Kraftstoff)	642 kg

Fahrbereich	440 kg
Nutzlast	450 kg

Baumuster K 1 (Typ 82)

Technische Daten

Motor

Arbeitsverfahren	4 Takt
Hub	64 mm
Bohrung	70 mm
Zylinderzahl	4
Hubraum (gesamt)	985 ccm
Verdichtungsverhältnis	1 : 5,8
Dauerleistung	23,5 PS
Drehzahl bei Dauerleistung	3000 U/min
Höchstleistung	24,5 PS
Drehzahl bei Höchstleistung	3300 U/min
Art der Kühlung	Luftkühlung
Ventilspiel (Einlaß u. Auslaß in kaltem Zustand)	0,12 bis 0,15 mm
Zündungseinstellung	5 Grad v. o. Totpunkt
Zündfolge	1 - 4 - 3 - 2
Art der Zündung	Batteriezündung
Zündzeitpunktverteiler	automatisch
Schmierung	Druckumlaufschmierung
Ölkühlung	Röhrenkühler
Vergaser	Solex-Fallstromvergaser

Fahrgestell

Kupplung	Einscheiben-Kupplung Fichtel und Sachs Größe K 10
Getriebe	Rädergetriebe
Zahl der Gänge	4 Vorwärtsgänge 1 Rückwärtsgang

Untersetzungen

1. Gang	2. Gang	3. Gang	4. Gang	R. Gang
1 : 3,6	1 : 2,07	1 : 1,25	1 : 0,8	1 : 6,6

Geschwindigkeit bei n = 3300 U/min	17 31 51 80 9 km/st
Hinterachsenantrieb	Kegelräderzähnezahl 7 : 31 Untersetzung 1 : 4,43
Zusatzgetriebe	je 1 an den seitlichen Hinterachswellen
Zusatzgetriebe	Stirnräder Zähnezahl 15 : 21
Zusatzgetriebe	Untersetzung 1 : 2,4
Antriebsachse	1 geteilte Pendelachse, hinten
Art der Federn	vorn und hinten Drehstäbe
Stoßdämpfer	hydraulisch; vorn einfach hinten doppelt wirkend
Bremsen	mech. Innenbackenbremse mit Seilzug
Fußbremse wirkt	auf 4 Räder
Handbremse wirkt	auf 4 Räder
Räder	Scheibenräder
Felgenart	Tiefbett
Einpreßtiefe	33 mm
Lochkreisdurchmesser	205 mm
Anzahl der Löcher	5
Bereifung	5,25 - 16 Gelände

Luftdruck	Vollast	Halblast
vorn	1,4 atü	1,2 atü
hinten	1,8 atü	1,5 atü

Lenkung	Sonderspindellenkung mit geteilter Spurstange Konstruktion Dr. Porsche
Sturz belastet 0 mm	leer 3 mm
Vorspur	3-6 mm
Nachlauf	2½ Grad
Spreizung	5 Grad
Radstand	2400 mm
Spurweite vorn	1356 mm
Spurweite hinten	1360 mm

Fahrzeug

Länge über alles	3740 mm
Breite über alles	1600 mm

Höhe über alles	Dach hochgeklappt 1650 mm	Dach niedergeklappt 1111 mm
Überhang	vorn 60 Grad	hinten 38 Grad

Bodenfreiheit	290 mm
Bauchfreiheit	220 mm
Betriebsfertiges Eigengewicht nach der StVZO	668 kg
Leergewicht	725 kg
Zulässiges Gesamtgewicht	1160 kg
Nutzlast	450 kg

Achsdrücke	vorn	hinten
bei Leergewicht	280 kg	445kg
bei Gesamtgewicht	487 kg	673 kg

Niedrigste Dauergeschwindigkeit	3 km/h
Autobahngeschwindigkeit	80 km/h
Höchstgeschwindigkeit	80 km/h
Steigvermögen auf losem Sand	40%
Steigvermögen auf der Straße	45%
Watvermögen	450 mm
Größte Zugkraft auf der Straße	33 kg
Größte Zugkraft auf dem Acker	552 kg
Kraftstoff-Normverbrauch	8 l/100 km

Berlin, 24. Februar 1940: Vorderansicht eines der ersten VW-Kübelwagen 82 Modelle (24 PS).

Berlin, 24. Februar 1940: Rückansicht eines der ersten VW-Kübelwagen 82 Modelle (24 PS).

Vorderer Kastenteil des VW-82: Unterhalb der heizbaren Windschutzscheibe die beiden Haltefedern. Links die Verschlußkappe des Kraftstoff-Einfüllstutzens und das Ersatzrad mit Deckkappe. Rechts oben der Winker, darunter Rückblickspiegel, Hupe und Tarnscheinwerfer Nova 1-1.01. Zu beiden Seiten des Mittelteils das auf den Zughaken befestigte Schleppseil.

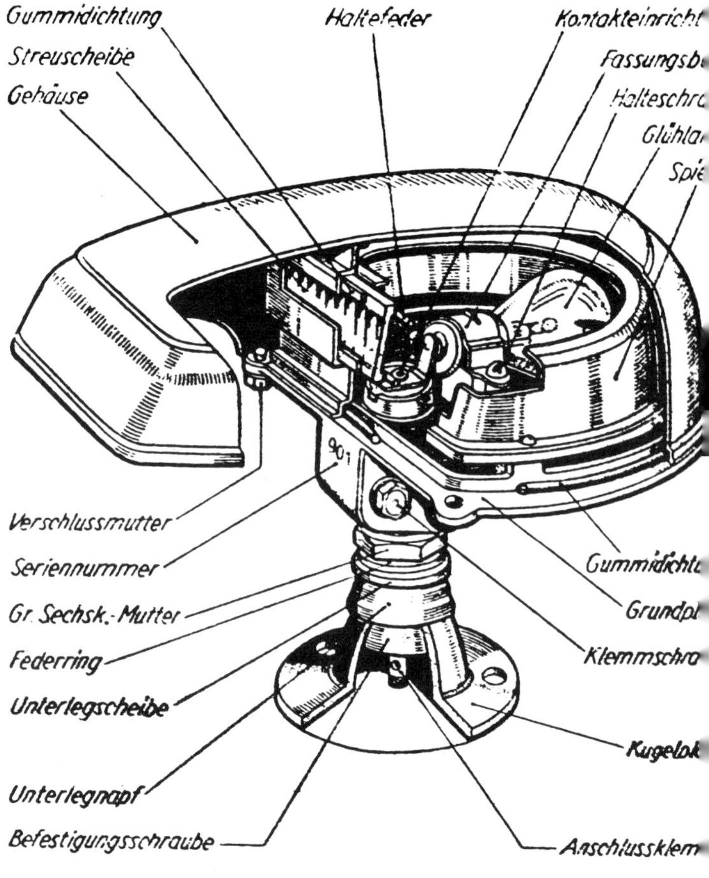

Der Tarnscheinwerfer im Detail.

Fahrbereich	400 - 450 km
Füllmengen	
Kraftstoffbehälter	30 l
Kraftstoffbehälter (Vorrat)	10 l
Kurbelgehäuse bei Ölwechsel	2,5 l Motorenöl
bei Zusammenbau d. Motors	3 l Motorenöl
Getriebegehäuse bei Ölwechsel	2,5 l Getriebeöl
Getriebegehäuse bei Zusammenbau d. Getriebes	3 l Getriebeöl
Lenkgehäuse	0,25 l Getriebeöl
Vorderachse	1,15 kg Abschmierfett

Gerätbeschreibung

Motor

Motorgehäuse
Das zweiteilige Motorgehäuse besteht aus Leichtmetall. Beide Hälften sind zusammen bearbeitet und dürfen nur zusammen ausgewechselt werden. Die 4 Zylinder aus Grauguß können einzeln ersetzt werden. Je zwei Zylinder tragen einen gemeinsamen abnehmbaren Zylinderkopf aus Silumin. Im Zylinderkopf liegen die Ventile und die Schwinghebel. Zwischen Zylinder und Zylinderkopf befindet sich keine besondere Dichtung.

Kurbeltrieb
Die 4 Hauptlager und die 4 Pleuellager der Kurbelwelle sind Gleitlager. Das zweite Hauptlager ist geteilt. Die Lagerstellen der Kurbelwelle sind gehärtet. Die Kolben sind aus Leichtmetall und tragen je einen konischen Ring oben, einen Kolbenring in der Mitte und einen Ölabstreifring unten. Alle Ringe sind gerade geschlitzt. Die Kolbenbolzen sind schwimmend gelagert und durch Federringe gesichert.

Steuerung
Die Nockenwelle wird durch zwei schrägverzahnte Steuerräder von der Kurbelwelle angetrieben. Das Nockenwellenrad ist aus Elektron. Die Steuerung der Ventile erfolgt durch Stößelstangen und Schwinghebel. Jeder Zylinder besitzt ein Einlaßventil und ein Auslaßventil.

Kühlung
Die Kühlung erfolgt durch Luft. Auf gleicher Welle mit der Lichtmaschine sitzt ein Gebläserad, das durch einen nachstellbaren Keilriemen von der Kurbelwelle angetrieben wird. Das Gebläserad saugt durch eine Öffnung im Gebläsegehäuse Luft an und preßt sie über die Zylinder. Die erforderliche Führung der Luft erfolgt durch Leitstücke, die im Gebläsegehäuse sitzen. Zwei weitere Leitstücke verkleiden die Zylinder.

Ölkühlung
Der Ölkühler sitzt im Gebläsegehäuse. Er ist in die Ölleitung eingeschaltet. Der Ölkühler besteht aus Röhren, die das Öl bei seinem Kreislauf dauernd durchfließen muß. Zwischen den Röhren wird die vom Gebläse angesaugte Kühlluft durchgepreßt. So wird das Öl auf einer hinreichend niedrigen Temperatur gehalten, die seine volle Schmierfähigkeit auch bei Höchstleistung des Motors gewährleistet. Ein Öldruckventil schaltet bei hohem Druck (kaltem Öl) den Ölkühler automatisch ab und läßt das Öl unmittelbar in die Leitungen fließen. In die Druckleitung zwischen Ölpumpe und Ölkühler ist ein selbsttätiger Schalter für die Öldruckprüflampe eingebaut, der bei einem Druck von 0,5–0,9 atü einen elektrischen Kontakt öffnet und dadurch den Strom der Öldruckprüflampe unterbricht.

Schmierung
Druckumlaufschmierung. Am Ende der Nockenwelle sitzt eine Ölpumpe, die den Ölkreislauf im Fluß hält. Das Schmieröl wird vom tiefsten Punkte des Motorgehäuses durch ein Rohr abgesaugt und zuerst durch den Ölkühler gedrückt. Hinter diesem verzweigen sich die Kanäle. Ein Teil des Öles wird durch die durchbohrte Kurbelwelle gepreßt, schmiert die Lager und tritt nachher wieder in das Motorgehäuse ein. Ein zweiter Teil fließt durch die Lager der Nockenwelle, schmiert diese und tritt ebenfalls in das Motorgehäuse ein. Ein dritter

Teil erreicht die Ölkanäle in den hohlen Stößelstangen, tritt in die Ölkanäle der Schwinghebel ein, schmiert deren Lager, rinnt durch die Schutzrohre der Stößelstangen wieder in das Motorgehäuse zurück. Die Zylinderwände, die Kolben und Kolbenbolzen werden durch Schleuderöl vom Motorgehäuse aus geschmiert. Die Ventilschäfte erhalten Öldunst und kleine Mengen Öl, das durch die Gewindeluft der Stellschraube der Schwinghebel auf den Ventilschaft übertritt. Die Ölreinigung erfolgt durch ein Sieb, das an der tiefsten Stelle des Motorgehäuses liegt.

Kraftstoffpumpe
Die mechanische Kraftstoffpumpe wird von der Zündverteilerwelle durch einen Nocken über einen Winkelhebel betätigt. Die Membran wird von keinem der üblichen Kraftstoffe chemisch angegriffen. Zur Abschirmung der Wärme des Motorgehäuses ist ein isolierender Zwischenflansch aus Preßstoff angeordnet.

Luftfilter
Das Luftfilter verhindert, daß mit der angesaugten Verbrennungsluft Staub in die Zylinder kommt. Im Filter wird durch die angesaugte Luft eine genau bemessene Ölmenge in Wirbelung gebracht und dadurch das eingebaute Sieb ständig benetzt. Eine einwandfreie Luftreinigung ist nur gewährleistet, wenn das Öl im Filter bis zur Anzeigeschraube steht.

Vergaser
Der Solex-Fallstromvergaser besitzt eine Luftklappe zum Anlassen. Beim Betätigen des Luftklappenzuges öffnet sich auch die Drosselklappe soweit, daß ein sicheres Anspringen gewährleistet ist. Das Saugrohr liegt außerhalb des Gebläsegehäuses. Die Gabelung des Saugrohres ist von einer Vorwärmekammer umgeben, durch die ein Teil der Auspuffgase geleitet wird. Die Auspuffgase treten durch ein Rohr in die Kammer. Hierdurch wird das angesaugte Gemisch vorgewärmt.

Elektrische Ausrüstung des Motors
Es ist eine Batteriezündung mit selbsttätiger Zündzeitpunktverstellung eingebaut. Der Sammler hat eine Spannung von 6 Volt und eine Kapazität von 75 Ampèrestunden. Die Lichtmaschine ist spannungsregelnd. Die Zündspule sitzt rückwärts am Gebläsegehäuse. Die Zündfolge ist 1–4–3–2. Die Zündeinstellung beträgt 5 Grad vor dem oberen Totpunkt. Der Anlasser ist ein Anlasser mit Zugmagnet.

Kraftübertragung

Kupplung
Einscheiben-Kupplung, Fabrikat Fichtel und Sachs, Größe K 10. Die Betätigung der Kupplung erfolgt vom Fußhebel aus durch ein Drahtseil, das im Mittelträger des Rahmens zum Kupplungshebel führt. Die Nachstellung erfolgt durch Muttern am Kupplungshebel.

Getriebe
Schaltgetriebe, Hinterachsantrieb und Ausgleichgetriebe sind in einem Gehäuse aus Leichtmetall vereint und am Motorgehäuse angeflanscht. Das Getriebe hat 4 Vorwärtsgänge und einen Rückwärtsgang mit Sperre. Der vierte Gang ist als Schnellgang ausgebildet. Erster, zweiter und Rückwärtsgang sind gerade verzahnt und haben Schubschaltung. Dritter und vierter Gang sind geräuscharm, schrägverzahnt und haben Stiftschaltung. Zum Schalthebel führt im Mittelträger des Rahmens die Schaltschubstange.

Hinterachsantrieb und Ausgleichgetriebe
Das Kegelrad hat 7 und das Tellerrad 31 Zähne. Die Untersetzung beträgt 1 : 4,43. Das Ausgleichgetriebe ist selbsthemmend, um in hohem Maße das Einzeldurchgleiten der Hinterräder auf glattem oder lockerem Boden zu verhindern. Es kommen die Fabrikate ZF oder Rheinmetall zur Anwendung.

Zusatzgetriebe
In einem Gehäuse an den seitlichen Achswellen stehen zwei Stirnräder, Zähnezahl 15 und 21 im Eingriff. Das ergibt eine Untersetzung des Antriebes von 1 : 1,4. Dadurch erhöht sich die Gelände- und Steigfähigkeit des Fahrzeuges. Die Anordnung des Untersetzungsgetriebes ergibt auch eine Höherlegung der Hinterachsen und somit eine größere Bodenfreiheit.

Fahrwerk

Achsen
Die Vorderachse besteht aus zwei miteinander starr verbundenen Rohren, die mit dem Rahmenkopf fest verschraubt sind. In den Achsrohren sind die Traghebel drehbar gelagert. An ihren Enden sitzen verstellbar, aber fest verschraubt, die Bundbolzen, welche die Lagerbügel tragen. Auf den Achsschenkelbolzen sind die Achsschenkel drehbar gelagert, an denen die Bremsträgerbleche und die Radlager sitzen.
Die Federung der Vorderräder ist eine Einzelradfederung durch Drehstäbe, die in den Achsrohren liegen. Sie bestehen aus 4 einzelnen Bandstahlstreifen von der ganzen Länge der Vorderachsrohre. In der Mitte sind die Stäbe durch eine Nuß und eine Feststellschraube gegen Verdrehung und seitliche Verschiebung gesichert. Die linke Hälfte der Drehstäbe dient als Abfederung des linken, die rechte Hälfte gilt als Abfederung des rechten Vorderrades. Die Enden der Stäbe sind verschweißt und greifen mit einem Vierkant in die Traghebel ein. Jeder Stoß auf die Vorderräder überträgt sich über die Traghebel auf die Drehstabfedern und verdreht sie bis zur Nuß. Anschläge mit Gummipuffern verhüten eine zu starke Durchfederung und hydraulische, einfachwirkende Stoßdämpfer verhindern ein Nachschwingen. Die Hinterachse ist als Pendelachse ausgebildet. In den Achsrohren drehen sich die Antriebswellen, die über das Zusatzgetriebe die Antriebskraft auf die Nabe der Hinterräder übertragen. Die Federung der Hinterräder ist eine Einzelradfederung und wird durch Drehstabfedern, die in einem Querrohr liegen, erreicht. Die Drehstäbe sind aus Rundstahl, sie reichen nur über die halbe Wagenbreite. Das innere Ende der Stäbe ist in einem Gußstück gelagert und verzahnt. Das äußere Ende sitzt ebenfalls mittels einer Verzahnung in der Nabe der Hinterachsstrebe. Das andere Ende der Strebe ist am Hinterachsrohr starr befestigt. Ein Stoß des Hinterrades überträgt sich über die Strebe auf die Drehstäbe und verdreht diese etwas. Die Stäbe sind mit Hilfe der Verzahnung verstellbar. Doppelwirkende hydraulische Stoßdämpfer verhindern ein Nachschwingen.

Räder
5 Scheibenräder je 5 Bolzen, Tiefbettfelgen 3,00 D–16.

Bereifung
Niederdruckreifen 5,25–16 Gelände.

Fahrgestellrahmen
Der Mittelträger ist aus Stahlblech zu einem halb elliptischen Querschnitt gepreßt und un-

ten mit einem angeschweißten Blech zu einem Tunnel abgeschlossen. Hinten teilt er sich in eine Gabelung zur Aufnahme von Motor und Getriebe. Vorn verbreitert er sich zum Rahmenkopf, der die Vorderachse trägt. Das Fußbodenblech ist angeschweißt. Durch den Mittelträger laufen in einem eingeschweißten Rohrbündel Drosselklappen-, Kupplungs-, Luftklappen- und die zwei hinteren Bremsseile, Druckstange der Feststellbremse und die Schubstange für die Getriebeschaltung. Die Kraftstoffleitung liegt ebenfalls im Mittelträger.

Lenkung
Die Übertragung der Lenkbewegung vom Lenkrad auf die Vorderräder geschieht in nachstehender Reihenfolge: Lenksäule, Gelenkscheibe, Lenkspindel, Lenkmutter, Lenkhebelwelle, Lenkhebel, Spurstangen, Achsschenkel, Räder. In den Lenkhebel greifen mit Kugelgelenken die linke und die rechte Spurstange ein. Die anderen Enden der Spurstange sind wieder durch Kugelgelenke mit dem rechten und linken Lenkhebel am Achsschenkel verbunden. Die Drehung der Achsschenkel erfolgt um die Achsschenkelbolzen. Verstellbar sind Lenkspindel und Lenkmutter.

Bremsen

Hauptbremse
Die Übertragung des Bremsdruckes vom Fußhebel erfolgt durch die Bremsschiene, die im Mittelträger des Rahmens nach vorne gedrückt wird. An ihrer Kopfplatte sind die Bremsseile mit Nippeln befestigt. Die Seile für die Vorderradbremse treten vorne aus dem Rahmen aus. Die Seile für die Hinterradbremse laufen durch den Mittelträger nach hinten und treten dort aus. Die Anpressung der Bremsbacken an die Bremstrommeln erfolgt durch ein Hebelwerk. Die Fußbremse wirkt auf alle 4 Räder.

Die Feststellbremse
wird durch einen Zug am Handhebel nach oben betätigt, der durch ein Zahnsegment feststellbar ist. Der Handbremshebel drückt die Bremsschiene mittels einer Druckstange nach vorn. Von der Schiene aus erfolgt die weitere Kraftübertragung wie bei der Hauptbremse auf alle 4 Räder. Hauptbremse und Feststellbremse sind also bis auf das Bedienungsorgan (Fuß- oder Handhebel) gleich. Die Bremsbacken können durch eine Stellschraube am Bremsträgerblech nachgestellt werden.

Hand- und Fußhebelwerk
Auf einem gemeinsamen Lagerbock sitzen Kupplungshebel, Fußbremshebel und Gashebel. Am Mittelträger des Rahmens Handbremshebel, Schalthebel und der Knopf für den Luftklappenzug.

Kraftstoffanlage
Der Kraftstoffbehälter ist vor dem rechten Vordersitz am Aufbau mit Spannbändern befestigt. Der Kraftstoffhahn ist als Dreiwegehahn ausgebildet. Sein Hebel kann drei Stellungen einnehmen: A = auf, Z = zu, R = Reserve. Der Hahn enthält das Kraftstoffilter mit dem Wasserabscheider. Ein biegsamer Schlauch führt zum Kraftstoffrohr, das im Mittelträger des Rahmens nach hinten führt. Ein zweiter Schlauch verbindet das andere Ende des Rohres mit der Kraftstoffpumpe. Die biegsamen Rohre sind mit dem Kraftstoffrohr im Mittelträger des Rahmens nicht verschraubt, sondern nur aufgeschoben.

Schaltbrett und Bedienungshebel
Am Schaltbrett sind angeordnet: Zündschloß, Anlasser-Druckknopf, Ladeprüflampe, Öldruckprüflampe, Anzeigelampe für Winker, Anzeigelampe für Fernlicht, Geschwindigkeitsmesser mit Kilometerzähler, Lichtschalter, Winkerschalter, Sicherungskästen, Steckdose für Handlampe, Mehrfachschalter für Nachtmarschgerät, Schalter für Schaltbrettleuchte.
Am Windschutz befinden sich 2 Scheibenwischer mit ihren Schaltern. Der Sucher hat einen Drehschalter. Am Lenkrad befindet sich der Knopf für das Signalhorn.
An Bedienungshebeln sind vorhanden: Kupplungshebel, Fußbremshebel, Gashebel, Zug für Vergaserluftklappe, Schalthebel, Handbremshebel, Kraftstoffhahn mit Filter, Fußschalter für Fernlicht.

Elektrische Anlage

Sammler
Der Sammler ist unter einem der Rücksitze untergebracht, er hat eine Spannung von 6 Volt und eine Kapazität von 75 Amperestunden.

Stromverbraucher
Zündung siehe Motor. Die Lichtanlage hat folgende Leuchtstellen: Scheinwerfer mit Fern-, Abblend- und Standlicht. Tarnscheinwerfer, Abstands-Rücklicht, Handlampe, Sucher, Warnlichter am Schaltbrett, Winkerlampen und zusätzliches zweites Schlußlicht. Signaleinrichtungen sind: Signalhorn und Winker.
Sonstige Stromverbraucher sind: Zwei Scheibenwischer mit Einzelantrieb und der Anlasser.

Sicherungen
Die Sicherungskästen liegen links und rechts am Schaltbrett und auf der linken Seite des Motorraumes. Die Zugehörigkeit der Sicherungen zu den einzelnen Verbrauchern geht aus dem Schaltschema und aus den Bezeichnungen an der Innenseite der Sicherungsdeckel hervor.

Entstörung
Die Kfz. sind teilentstört nach Gruppe III. Ein Teil der Kfz. ist vollentstört nach Gruppe I.

Stoßdämpfer
Das Fahrzeug ist mit hydraulisch wirkenden Stoßdämpfern ausgerüstet, die ein Nachschwingen beim Federn verhindern. Die hinteren Stoßdämpfer sind doppelt, die vorderen einfach wirkend. Die vorderen dürfen nicht zerlegt werden. Auch die Gummimanschette darf nicht entfernt werden.

Aufbau
Der aus Ganzstahl ausgeführte Aufbau hat je einen Geräteraum über der Vorder- und Hinterachse; die Türen sind an der Mittelsäule fest angelenkt, die Kotschutzbleche mit dem Aufbau verschraubt. Die Windschutzscheibe ist aus Sicherheitsglas und nach vorne umklappbar. Es sind Steckfenster vorhanden, die bei Nichtgebrauch in besonderen Abteilen gelagert sind. Die Vordersitze mit fester Rücklehne am Fahrgestell sind verstellbar. Die Hintersitzbank ist in die Karosserie eingelegt, die hintere Rücklehne bei geöffnetem Geräteraumdeckel vorklappbar.

Unterbringung von Werkzeugen und Zubehör
Der Zeichenstab ist vor dem rechten Vordersitz und der Verbandkasten links unter der Windschutzscheibe untergebracht.
Links vorne an der Seitenwand ist ein Kasten

für die Wagenpapiere und die Handlampe angebracht.
Der Geräteraum für Nachtmarschgerät und Fahrersturmgepäck ist vorne im Wagen, Sturmgepäck für Beifahrer wird teils unter den beiden Vordersitzen und teils rechts unter dem Hintersitz verpackt. Die Schneeketten befinden sich rechts neben der Batterie unter dem Hintersitz.
Das Werkzeug und der Wagenheber sind im Motorraum untergebracht. Die Gewehrhalter befinden sich an der Haltestange über den Vordersitzen und die Gewehrschuhe sind am Bodenblech befestigt.
Der Spaten ist an der rechten Seitenwand und dem Vorderkotflügel in Halterungen gelagert.

Geschwindigkeitsmesser
Er wird durch zwei Kegelräder mit Dauerschmierung im linken Achsschenkel angetrieben.

Viele der normalen VW-Kübelwagen 82 wurden serienmäßig im Jahre 1942–1945 mit verschiedenen Gasgeneratoren ausgerüstet. Diese Entwicklung lief unter der Typenbezeichnung »230«, »231« und »239«. Hier der VW-Kübelwagen »239« mit Holzkohlengenerator.

Reichsgebiet, Umgebung von Stuttgart, Mai 1941: Der gute VW-Kübelwagen in einem neuen Kleid, diesmal als Panzerkampfwagen-Attrappe mit der Bezeichnung Porsche Typ 82/3.

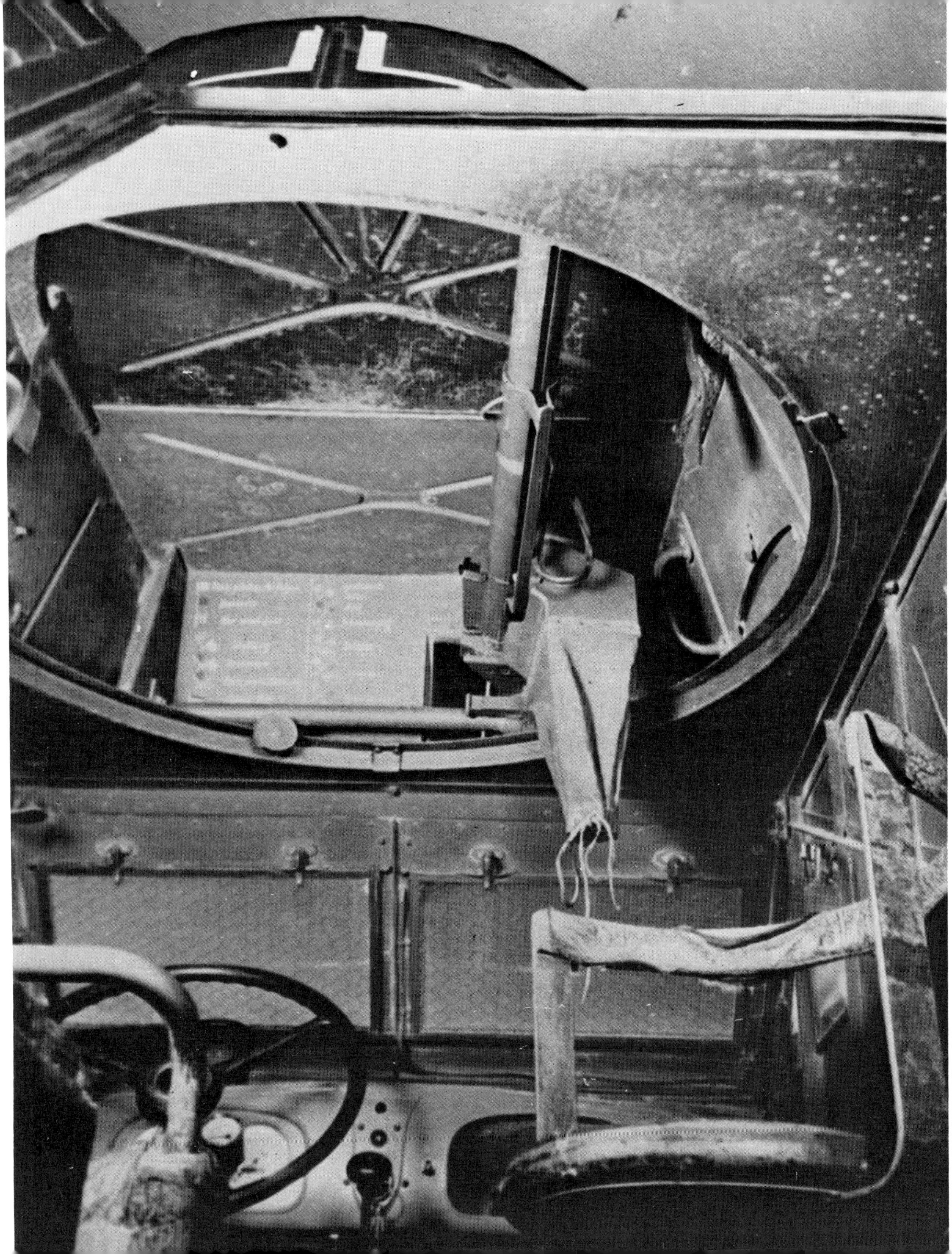

Porsche Typ 82/3: Ohne Radverkleidung mit dem MG-Turm. Die Einstiegluke befindet sich an der rückwärtigen Seite.

◀ Porsche Typ 82/3: Der Kfz-Führer- und Beifahrersitz (von hinten schräg nach oben aufgenommen) mit dem Drehturm für das MG 34.

Wozu ist der VW-Kübelwagen noch gut: Der Porsche Typ 82/3 wird meistens nur noch zu Ausbildungszwecken verwendet.

Technische Angaben

Hauptlager	Axialspiel 0,10 - 0,20 mm Radialspiel im 1. - 3. Lager 0,05 - 0,07 mm im 4. Lager 0,03 - 0,05 mm Das Axialspiel wird am ersten Hauptlager gemessen. Beachte die Druckscheibe 0,98 mm.
Kolben	Passung 0,05 - 0,07 mm am Kolbenmantel unten in Drehrichtung gemessen. Zum Einbringen des Bolzens Kolben gut handwarm erwärmen.
Kolbenringe	Stoßluft: Erster Kompressionsring 0,20 bis 0,35 mm Zweiter Kompressionsring 0,20 bis 0,35 mm Ölabstreifung 0,20 - 0,35 mm Der erste Kompressionsring ist konisch. Beachte die Bezeichnung Top oder oben. Kolbenring-Nutenspiel 0,03 bis 0,05 mm
Kompressionsdruck	Bei Anlasserbetätigung und warmer Maschine 4 - 5 atü. Zündkerzen herausnehmen.
Kraftstoffpumpe	Pumpendruck 1 - 1,2 m. W. S. Vorspannung des Pumpenstößels 0,5 - 1,0 mm
Kupplung	Das Kupplungsfußhebelspiel beträgt 20 - 30 mm. Es wird am Seilzugende, das zum Kupplungshebel außen am Triebwerkgehäuse führt, eingestellt. Die Betätigungshebel für die Kupplungsanpreßplatte werden, nachdem sie vom Werk eingestellt worden sind, nicht wieder verstellt.
Lenkung	Wenn sich Kleben oder Schwergehen der Lenkung bei Befahren glatter, ebener Straßen einstellt, dürfte die Kupferplattierung der Lenkmutter abgenützt sein, alsdann muß die Lenkmutter ausgewechselt werden.
Nockenwelle	Lagerspiel: Achsialspiel 0,02 - 0,07 mm Radialspiel 0,02 -0,05 mm
Öldruck-Kontrollschalter	Öffnet bei atü 0,5 - 0,9
Öldruckventil	Die Länge der unbelasteten Feder beträgt 45,7 mm
Ölfüllung des Kurbelgehäuses	Die Neufüllung des Kurbelgehäuses beträgt 3 Liter. Bei Ölwechsel beträgt die Füllung 2½ Liter, weil ½ Liter im Ölkühler zurückbleibt. Wegen des Zurückbleibens von Öl darf auch kein Spülöl verwendet werden.
Ölfüllung des Triebwerkgehäuses	Die Neufüllung des Triebwerkgehäuses beträgt 3 Liter. Bei Ölwechsel beträgt die Füllung 2½ Liter
Pleuel	Lagerspiel: Axialspiel 0,1 - 0,2 mm Radialspiel 0,035 - 0,065 mm
Stirnräder	Zahnflankenspiel 0,01 - 0,035 mm
Stößel	Um Stößelgeräusche zu vermeiden, halte das Spiel zwischen Stößel und Führungsblech womöglich unter 0,01 mm. Spiel Stößelbohrung 0,02 bis 0,05 mm
Teller- und Kegelrad	Zahnflankenspiel: Klingelnberg 0,1 - 0,15 mm Cleason 0,1 - 0,12 mm
Triebwerk	Preßsitz des druckaufnehmenden Kugellagers der Kegelradwelle 0,02 - 0,05 mm
Ventile	Ventil-Stößelspiel bei kalter Maschine 0,15 mm. Benutze Stufenfühlerlehre VW 142 Ventilschaft-Führungsspiel: E: 0,04 - 0,06 mm A: 0,05 - 0,07 mm
Ventilsitzring	Sitzflächenwinkel 45° Sitzflächenbreite 1,4 mm
Verteilerantrieb	Zahnflankenspiel 0,1 - 0,2 mm
Verteiler	Unterbrecherkontakt-Abstand 0,4 - 0,5 mm
Vorderachse	die Achsneigung beträgt 2° 30'. Das untere Tragrohr sitzt also dem oberen um 2°30' = 6 mm voraus. Die Achsneigung kann nicht verändert werden! Der Sturz zum Rad war an der Felge oben und unten gemessen 3 - 5 mm. Beachte: der richtige Sturz wird bestimmt durch eine Stahlscheibenbeilage von 0,2 mm und einer Preßstoffscheibenbeilage von 4,5 mm, die jedoch beide Verschleiß unterworfen sind! Die Vorspur beträgt 4 - 6 mm in der Höhe der Felgenmitte hinten und vorne gemessen
Zündkerzen	Elektroden-Abstand 0,6 mm

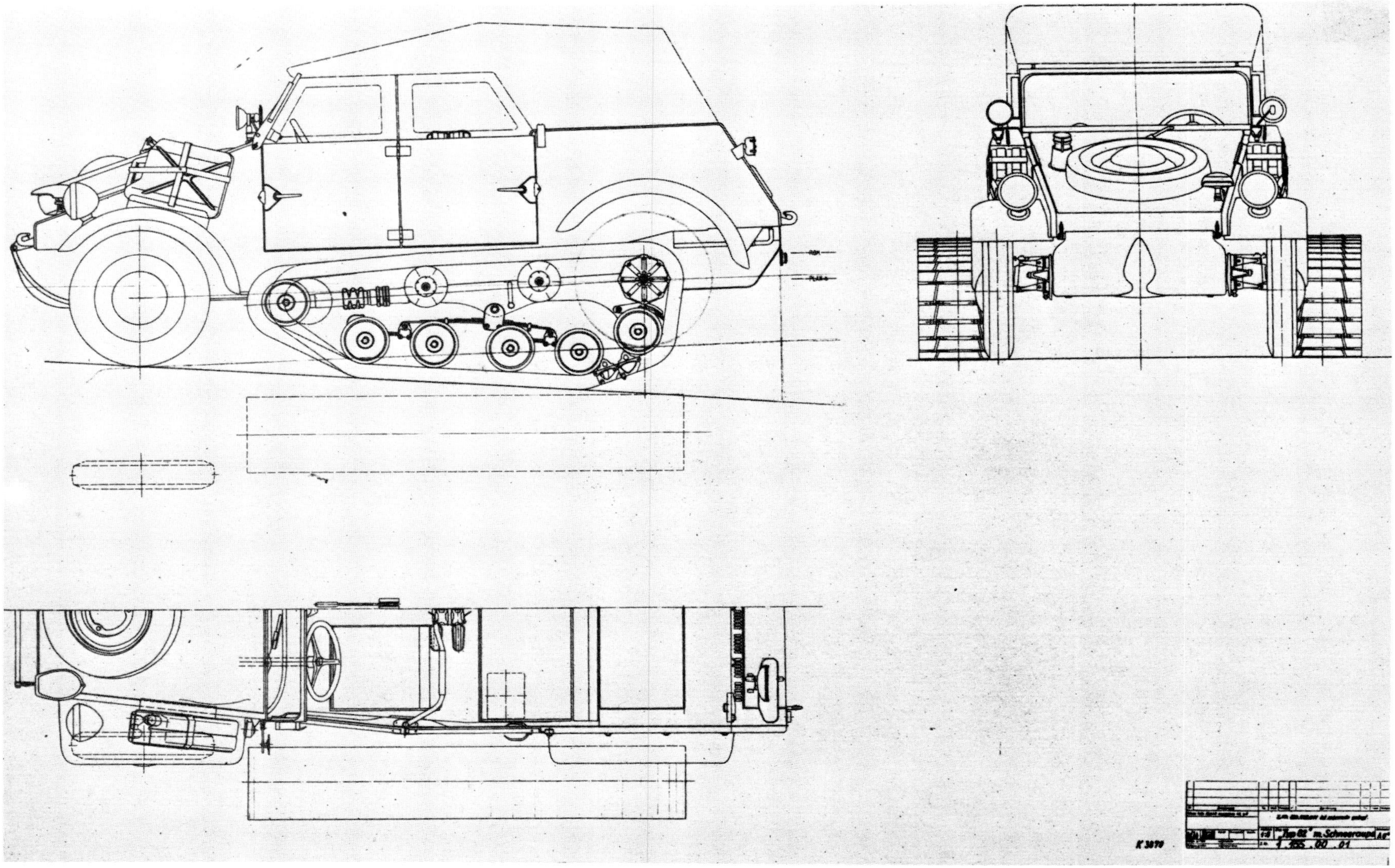

Der VW-82/Porsche Typ 155 mit Schneeraupe. Maßstab 1:29.

Reichsgebiet, Schwarzwald, Herbst 1942: Der VW-82/Porsche Typ 155 Schneeraupe I bei Gelände-Probefahrten. Dahinter ein NSU Ketten-Krad HK-101.

Stuttgart-Zuffenhausen, Herbst 1942: VW-Kübel/Porsche Typ 155 als Schneeraupe II.

Stuttgart-Zuffenhausen, Herbst 1942: VW-82/ Porsche Typ 155 Schneeraupe IV.

Stuttgart-Zuffenhausen, September 1942: Der VW-82 Kübelwagen als Halbkettenfahrzeug Porsche Typ 155. Der mittlere Teil der Schneeraupe III mit einem der vier Kettenlaufwerk-Systeme.

Verdunkelungsvorschriften

An den unbedingt verkehrsnotwendigen Aussenlichtquellen sind Verdunkelungsmaßnahmen durchzuführen. Die übrigen Lichtquellen sind außer Betrieb zu setzen.

Die Scheinwerfer sind so abzudecken, daß nur ein waagerechter, 5 bis 8 cm langer, 1,5 cm breiter Ausschnitt in der Mitte der Abschlußscheiben das Licht austreten läßt. Auch hier muß auf freier Landstraße mit Fernlicht, bei Gegenverkehr und geschlossenen Ortschaften mit Abblendlicht gefahren werden. Die übrigen Leuchten sind in ihrer Leuchtwirkung so herabzusetzen, daß sie nur auf eine Entfernung bis zu 500 m wahrnehmbar sind. Schlußleuchte und Kennzeichenleuchte lassen sich leicht dadurch abdunkeln, daß man statt der bisherigen 6 V-Lampen solche zu 12 oder 24 V nimmt. Bei Innenbeleuchtung aller Art sind die Stellen, an denen Licht austreten kann, abzublenden.

Um auch bei Verdunkelung die Betriebssicherheit im Kraftfahrzeugverkehr nicht zu gefährden, mußten aufgrund des Gesetzes über den Verkehr mit Kraftfahrzeugen ab 1. Oktober 1940 die Fahrzeuge mit dem Wehrmacht-Tarnscheinwerfer ausgerüstet werden. Dieser ist schwingungsfrei so anzubringen, daß die Lichtaustrittsöffnung nicht höher als 1,2 m und nicht tiefer als 0,8 m über der Fahrbahn liegt. Er ist so zu neigen, daß die obere Hell-Dunkel-Grenze des Lichtes in einer Entfernung von 5 m vor der Lichtaustrittsöffnung mindestens 6 cm tiefer als die obere Kante der Lichtaustrittsöffnung liegt. Er muß ferner unabhängig von der übrigen Beleuchtung auszuschalten und so geschaltet sein, daß er nur mit den Begrenzungsleuchten zugleich brennen kann. Der Lichtfächer wird flach üb r die Fahrbahn vorauseilend und breit hingestreut. Von der Luft aus ist er schwer festzustellen, von einer gewissen Höhe aus wird er unsichtbar. Der Tarnscheinwerfer ist völlig blendfrei. Die Kugelplatte am Fuß ermöglicht einfachen Anbau und einfache Einstellung.

Leichter Personenkraftwagen K 1 Typ 82

Sonderausrüstung für Tropen

Allgemeines

Durch Einbau der Sonderausrüstung erhält das Kfz. einen zusätzlichen Schutz gegen Sand, Staub und Hitze. Ferner wird durch Einbau anderer Reifengrößen die Geländegängigkeit wesentlich erhöht. Zum Einbau gelangen folgende Teile:

1 Ölmeßstab mit Schutzschlauch,
1 Kraftstoffhilfspumpe mit Zusatzfilter,
2 Abdeckbleche für Saugrohrheizung,
1 Schutzkappe für Zündspule,
1 Schutzkappe für Zündverteiler,
1 Schutzkappe für Regler der Lichtmaschine,
1 Antriebswelle mit besonderem Krümmungsbogen für Geschwindigkeitsmesser,
5 Scheibenräder,
5 Reifen Größe 690 × 200,
5 Schläuche mit Gummiventil Größe 690 × 200,
5 Raddeckel,
1 Beilage für oberes Tragrohr der Vorderachse,
1 Zwischenstück für Ersatzradträger.

Durch den Einbau der Sonderausrüstung entfallen:

1 Ölmeßstab,
1 Antriebswelle für Geschwindigkeitsmesser,
5 Scheibenräder »bisherige Ausführung«,
5 Reifen,
5 Luftschläuche,
5 Raddeckel.

Sonderausrüstung des Motors

Ölmeßstab mit Schutzschlauch
Der verlängerte Ölmeßstab mit Schutzschlauch verhindert das Eindringen von Sand und Staub durch das Führungsrohr des Ölmeßstabes am Kurbelgehäuse.

Kraftstoffhilfspumpe
Die Kraftstoffhilfspumpe ist in die Kraftstoffleitung zusätzlich eingebaut. Wenn durch die vom Motor angetriebene Kraftstoffpumpe nicht genügend Kraftstoff zum Vergaser gefördert wird oder die Kraftstoffzuführung durch Gasblasenbildung unregelmäßig ist, dann kann durch Betätigen der Hilfspumpe schnell genügend Kraftstoff zum Vergaser gefördert werden. Desgleichen ist vor dem Anlassen ein Auffüllen des Schwimmergehäuses möglich, wenn durch Strahlungswärme Kraftstoff verdunstet ist.

Abdeckbleche für Saugrohrheizung
Durch Einlegen der Abdeckbleche wird die Saugrohrheizung ausgeschaltet, damit die bereits warme Außenluft nicht zusätzlich erhitzt wird.

Schutzkappen für elektrische Ausrüstung
Die Schutzkappe für Zündspule verhindert das Eindringen von Wasser in den Anschlußkopf der Zündspule.
Die Schutzkappe für den Zündverteiler verhindert das Eindringen von Sand und Staub, gewährleistet aber eine genügende Belüftung zum Vermeiden von Oxydation der Innenteile.
Die Schutzkappe zum Regler der Lichtmaschine dient dem gleichen Zweck.

Sonderausrüstung des Fahrgestells

Antriebswelle für Geschwindigkeitsmesser
Die Antriebswelle für den Geschwindigkeitsmesser hat einen besonderen Krümmungsbogen, damit die verbreiterte Radfelge den Geschwindigkeitsmesserantrieb nicht beschädigt.

Räder mit Bereifung
Durch die stark vergrößerte Bereifung, welche mit einem wesentlich niedrigeren Luftdruck gefahren werden kann, sinkt der spezifische Bodendruck erheblich. Das Fahren in losem Sand usw. wird dadurch stark erleichtert.

Beilage für oberes Tragrohr der Vorderachse
Die Sonderbereifung erfordert einen Nachlauf = O. Um den laut D 662/6 Abschnitt A vorhandenen Nachlauf von 2½° zu beseitigen, ist hinter das obere Tragrohr der Vorderachse ein Beilagstück eingebaut.

Zwischenstück für Ersatzradträger
Durch das Zwischenstück wird die Auflage des Ersatzradträgers so weit erhöht, daß das Ersatzrad mit der vergrößerten Bereifung untergebracht werden kann.

Beachte bei Betrieb des Kfz. Typ VW 82 in den Tropen:

Motoröl: Verwende dickes Motoröl von einer Viskosität von rund 25° Engler bei 50° C.

Abschmieren: Schmiernippel gut von Sand reinigen vor dem Abschmieren. Nach dem Abschmieren die Fettkragen an den Achsschenkelbolzen, Bundbolzen usw., ebenso die

Fettpolster auf den Schmiernippeln nicht entfernen, sondern zum Schutz gegen Sand und Staub sitzen lassen.

Luftfilter: Alle 500 km den Wirbelölfilter reinigen und neu mit Motoröl bis zur Kontrollschraube füllen. Die Klemmschellen des Rohres zwischen Filter und Vergaser stets so setzen, daß die Schlitze des Rohres verdeckt sind.

Bei Ölkontrolle vor dem Herausziehen des Meßstabes den Stab und den Schutzschlauch von Sand reinigen.
Beim Ölwechsel zuerst die nähere Umgebung der Ablaßschraube, ebenso den Einfüllstutzen von anhaftendem Sand reinigen.

Bei Getriebeölwechsel: Ablaßschraube und Einfüllstutzen vor dem Öffnen von Sand reinigen.
Schmierstoffbehälter, Schmierstofftrichter, Abschmiergerät vor ihrer Benutzung innen und außen gut reinigen, damit kein Sand mit dem Schmierstoff in die Innenteile des Fahrzeuges gelangt.
Kraftstoffgefäße, Kraftstofftrichter ebenso behandeln, damit kein Sand in den Kraftstoffbehälter gelangt. Kraftstoff stets durch ein Filtertuch nachfüllen.

Kraftstoffilter alle 1000 km reinigen: 1. Den kleinen Filter am Kraftstoffhahn: Filterkappe abnehmen, reinigen, Filtersieb mit Kraftstoff und Pinsel reinigen. 2. Zusatzfilter an der Hilfspumpe: Filterkappe abnehmen, reinigen und den Wollfiltereinsatz mit Kraftstoff und Pinsel reinigen.

Arbeiten am Motor: Bei allen Hantierungen am Motor peinlich sauber vorgehen, damit kein Sand in den Motor gelangt und seine Betriebssicherheit gefährdet.

Hilfspumpe nur betätigen, wenn die Kraftstoffzufuhr durch auftretende Gasblasenbildung behindert wird.

Keilriemenspannung kontrollieren alle 1000 km, zur Sicherung der Motorkühlung, wenn nötig, nachstellen.

Leichter Personenkraftwagen K 1 Typ 82

Sonderausrüstung für den Winterbetrieb

Bedienungs-Anweisung
Anlassen bei Temperaturen unter − 15°
während des Anlassens stets den Kupplungs-Fußhebel niedertreten. Vor der Anlasserbetätigung ist der Motor mit der Handkurbel etwa 20mal durchzudrehen. Die Starterklappe soll dabei geschlossen und die Zündung eingeschaltet sein. Ist während diesem Handandrehen noch keine Zündung erfolgt, so gieße man nun etwa 15–20 ccm Anlaßkraftstoff oder handwarmes Fahrbenzin in die Füllkapsel (½) und betätige nun den Anlasser. Liegt die Motordrehzahl sehr niedrig, so lasse man durch eine zweite Person mitkurbeln, wodurch dann der sehr schwache Sammler entlastet und damit der Zündfunken kräftiger wird. Ein Startversuch soll möglichst jeweils eine Minute lang durchgeführt werden. Erfolgen keine Zündungen, so kann weiterer Anlaßkraftstoff bis insgesamt etwa 150 ccm in Mengen von jeweils 30 ccm eingegossen werden. Ein Naßwerden der Zündkerzen ist dabei nicht zu vermeiden, bildet aber auch kein wesentliches Starthindernis. Die Drosselklappe soll während des Startvor-

ganges geschlossen, jedoch hin und wieder leicht geöffnet werden.
Springt der Motor an und bleibt aber nach kurzer Laufzeit wieder stehen, so ist meistens der Anlaßkraftstoff verarbeitet, der Motor jedoch noch nicht so warm, um aus dem normalen Fahrkraftstoff ein zündfähiges Gemisch bilden zu können. Dieser Zustand tritt besonders bei einer Kälte unter − 30° ein. Bei geringen Kältegraden (auch im Sommer) kann das Stehenbleiben des Motors auch auf ein überfettes Gemisch zurückzuführen sein.

Gegenmaßnahme
Drosselklappe auf Vollgasstellung bringen und Anlasser betätigen bzw. mit Handkurbel Motor einige Male durchdrehen.

Kälte unter − 30°
Das Anlassen bei diesen Kältegraden ist stets nur unter Benutzung des Anlaßkraftstoffbehälters durchzuführen. Da außerdem die Kraftstoffmembranpumpe bei diesen Temperaturen und den geringen Motordrehzahlen nicht oder nur sehr unvollkommen arbeitet, wird der zum Anlassen notwendige Kraftstoff nicht gefördert.

Volkswagen (Typ 82), schienengängig
Der Volkswagen (Typ 82), schienengängig (VW schg), ist kein »Nur-Schienenfahrzeug«, sondern das normale, straßen- und geländegängige Kraftfahrzeug mit einer zusätzlichen Schienenausrüstung. Der VW schg kann daher je nach örtlichen Verhältnissen wechselweise auf Straßen, im Gelände und auf Schienen eingesetzt werden. Die Schienengängigkeit macht die Verwendung des Fahrzeugs auch von jahreszeitlichen Witterungseinflüssen weitgehend unabhängig.
Die Sicherheit des VWschg gegen Entgleisungen ist aufgrund des geringen Fahrzeuggewichts kleiner als bei Eisenbahnfahrzeugen; es sind deshalb Beschränkungen der Fahrgeschwindigkeit im Schienenbetrieb erforderlich.
Das Fahrzeug läuft nicht mit besonderen Schienenrädern auf Schienen, sondern mit seinen luftbereiften Straßenrädern, welche zur Anpassung der Spurweiten an den Schienenabstand mit den Innenseiten nach außen gewendet sind. Das Abrutschen der Straßenräder von den Schienen wird durch neben den Rädern angeordnete Spurscheiben verhindert. Die Spurscheibe verhindert ein seitliches Abrutschen des Straßenrades von der Schiene; beide Spurscheiben laufen mit ihren äußeren Spurkranzflächen an den Schienen entlang und führen hierdurch die Achse. Beim Durchfahren von Weichen übernimmt der Spurkranz, welcher der Schienenlücke am Weichen-Herzstück gegenüberliegt, die Führung mit seiner Innenfläche am Radlenker. In Abhängigkeit von der Schienenspur beträgt der lichte Abstand zwischen den Spurkränzen in Schienenkopfhöhe 1360 ± 3 mm.
Die Spurscheibe besteht aus einer 3 mm starken Stahlblechscheibe, deren Außenrand spurkranzförmig gebördelt ist. Abweichend vom Laufprofil des Eisenbahnrades ist die Anlauffläche des Spurkranzes am Schienenkopf steiler ausgeführt; diese Formgebung ist mit Rücksicht auf die geringen Raddrücke des Fahrzeugs zur Vermeidung von Entgleisungen notwendig. Aus dem gleichen Grunde beträgt der Außendurchmesser der Spurscheibe 711 mm; dieses Maß ergibt sich aus der Bedingung, daß die für den Gleisbetrieb erforderliche Spurkranzhöhe von 27 mm auch bei dem nur mit dem Fahrer belasteten Fahrzeug mit entspre-

chend geringsten Reifeneinpressungen und gleichzeitig noch nicht abgenutzten Reifen vorhanden sein muß.
An der Innenseite der Spurscheibe sind drei Zentrierteile angeordnet, welche das Aufschieben und Festhalten der Spurscheibe beim gleichzeitigen Anschrauben von Spurscheibe und Straßenrad erleichtern.
Die Bohrungen für die Bundschrauben sind durch Kugelhülsen verstärkt; sie dienen außerdem zum Zentrieren des bei Schienenfahrt gewendeten Straßenrades.
Um bei dem gewendeten Rad das nach innen stehende Ventil zugänglich zu machen, ist die Spurscheibe mit einer runden Aussparung versehen; die gegenüberliegende Aussparung ist nur aus Gründen des Gewichtsausgleichs zur Vermeidung von Schwingungen angeordnet.
Der Schienenbetrieb ist unabhängig von der jeweiligen Fahrzeugbelastung durchführbar; jedoch ist notwendig, daß die vorgeschriebenen Reifendrücke vorhanden sind und daß sie sowohl bei den Vorderrädern, als auch bei den Hinterrädern unter sich genau übereinstimmen. Überlastungen des Fahrzeugs sind nicht zulässig.
Die Schienenausrüstung ist so leicht wie möglich gehalten und läßt sich bei Straßen- und Geländefahrten ohne Verlust an Nutzraum mitführen; die Größtnutzlast wird hierdurch nur um etwa 70 kg eingeschränkt. Das schienengängig ausgerüstete Fahrzeug kann daher – wie der normale Wagen – außer dem Fahren drei Mann mit Ausrüstung oder die entsprechende Nutzlast befördern.
Der Umbau des Fahrzeugs vom Straßen- zum Schienenfahrzustand – und umgekehrt – ist rasch durchführbar und kann notfalls auch unabhängig von Wegübergängen auf der freien Strecke vorgenommen werden. Das gleiche trifft für das Wenden des Wagens auf Schienen zu.

Ostfront, Abwehrkämpfe bei Kirowograd, Herbst 1943: Ein VW-Schwimmwagen Typ 166. Neben dem Fahrer Oberst Lorenz, Kommandeur des Panzergrenadier-Regiments »Großdeutschland«, hinter ihm Major Chrapkowski, Kdr. des Pionier-Bataillons »Großdeutschland«. Der VW-166, ein Bruder des VW-Kübelwagens, wurde von den Truppen seiner einmaligen Geländetüchtigkeit wegen geschätzt und in den Jahren 1942–1944 insgesamt 14 283 mal gebaut.

WAR DEPARTMENT TECHNICAL MANUAL

TM E9-803

GERMAN VOLKSWAGEN

WAR DEPARTMENT • *6 JUNE 1944*

RESTRICTED *DISSEMINATION OF RESTRICTED MATTER*— The information contained in restricted documents and the essential characteristics of restricted materiel may be given to any person known to be in the service of the United States and to persons of undoubted loyalty and discretion who are cooperating in Government work, but will not be communicated to the public or to the press except by authorized military public relations agencies. (See also paragraph 23, AR 380-5, 15 March 1944.)

Ein am Tage der Invasion in der Normandie, am 6. Juni 1944, für die kämpfenden Truppen vom US-Kriegsministerium herausgegebenes detailliertes Technisches Handbuch über den VW-Kübelwagen.

GERMAN VOLKSWAGEN

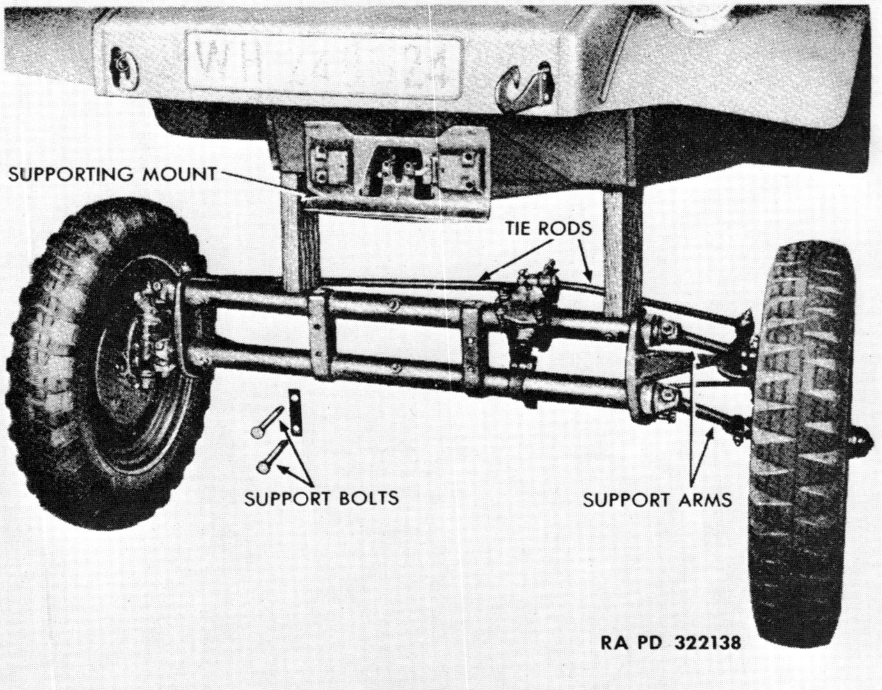

Figure 45—Front Suspension Removed

Eines der zahlreichen Bilder aus dem US-Kübelwagen-Handbuch.

»Eisenbahn-Transport eines VW-Kübelwagens« – aus dem US-Handbuch. ▶

SHIPMENT AND TEMPORARY STORAGE

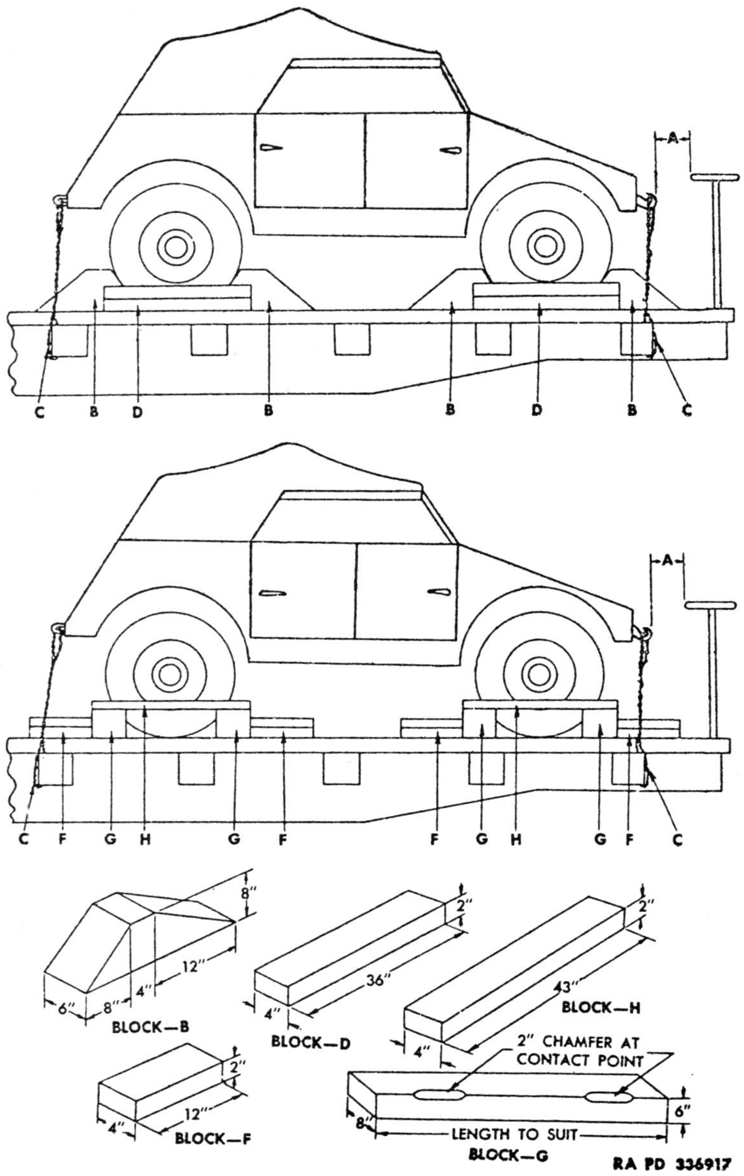

Figure 63—Blocking Requirements for Securing Vehicles on Railroad Cars

Typ 87 / 1942 - 1944

Motor
luftgekühlter 4-Takt-Ottovergasermotor mit Batteriezündung, liegende Bauart, gegenläufig, obengesteuert.

Bohrung	75 mm
Hub	64 mm
Hubraum	1150 cm^3
Verdichtung	1 : 5,8
Leistung	27,5 PS
Höchstdrehzahl	3300 U/min
Normaldrehzahl	3000 U/min
Höchstdrehmoment	760 cm/kg bei 2000 U/min
Kurbelgehäuse	Leichtmetall
Zylinder	Grauguß
Zylinderkopf	Silumin mit eingezogenem Ventilsatz
Ventile je Zylinder	1 Einlaß-, 1 Auslaß
Steuerung	Stoßstangen-Kipphebel, Nockenwelle untenliegend
Kurbelwelle	4 Gleitlager (Bleibronce)
Kolben	Leichtmetall
Gebläseantrieb über Lichtmaschine	durch Keilriemen
Drehzahl des Gebläses	6000 U/min
Luftmenge	500 l/min
Schmierung	Umlaufschmierung durch Zahnradpumpe
Ölkühler	Röhrenkühler
Zündfolge	1 - 4 - 3 - 2
Verteilerantrieb	Schraubenräder Übersetzung 2 : 1
Vergaser	Fallstromvergaser 26 ∅ Typ 26 VFJ (Deutsche Vergasergesellschaft)

Kupplung

Einscheiben	Trockenkupplung Größe K 10 mit doppelten, wärmeisolierten Federn (Fichtel und Sachs)

Getriebe

Gangzahl	vorwärts 5; rückwärts 1
Untersetzung	1. Gang = 1 : 3,60 2. Gang = 1 : 2,07 3. Gang = 1 : 1,25 4. Gang = 1 : 0,8 R. Gang = 1 : 6,6
3. und 4. Gang	geräuschlos Geländeg. 1 : 6
Zahnradausführung	1. u. 2. Gg. u. R. Gg. gerade verzahnt, Schubschaltung 3. u. 4. Gg. schräg verzahnt, Stiftschaltung Kugel
Art der Schaltung	Geländegang u. Vorderradantrieb Mit Hebel, Mitte Wagen

Vorderachse

Kegelräder	5 : 31
Ausgleich	(keine Spurdifferenz)

Hinterachse

Antrieb	Kegelräder 7 : 31
Übersetzungsgetriebe	Stirnräder außen bei den Rädern 15 : 21
Ausgleich	Z.F. selbsthemmend

Wagengeschwindigkeit

bei Motor n = 3300 U/mm	Geländeg. = 10 km/h 1. Gang = 17 km/h 2. Gang = 32 km/h 3. Gang = 51 km/h 4. Gang = 80 km/h R. Gang = 9 km/h

Steigfähigkeit

bei Motor n = 2000 U/mm	im Geländeg. = 75%

Fahrgestell

Rahmen
drehungssteifer Mittelrahmen mit Gabelung hinten zur Aufnahme von Motor mit Getriebe, Fußboden mit dem Rahmen fest verbunden. Träger des Vorderachsenantriebes und der Vorderachse mit dem Rahmen verflanscht.

Antriebsaggregat
Motor, Getriebe und Achsantrieb zu einem Block verflanscht und in 3 Punkten um Rahmen gummigelagert, Vorderachsantrieb auskuppelbar. Rohrwelle mit Gummigewebescheibe im Rahmen.

Federung
vorne unabhängig gefederte Räder über Stabfedern. Hinterräder, Schwingachse mit Strebenabstützung, einstellbarer Federstab (Kerbzahnprofil, Noniusteilung). Dämpfung über hydraulische Stoßdämpfer.

Lenkung
Sonderspindellenkung Konstruktion Dr. Porsche.
Lage der Lenkung links, Spurstange geteilt.

Übersetzung	2,4 Umdr. am Lenkrad = 49° = voller Einschlag der Vorderräder 1 Umdr. am Lenkrad = 20,5° = 42%
Lenkrad	400 mm ∅
Wendekreis	10 m ∅ gemessen über Kotflügel
Radeinschlag	äußeres Rad 28°, inneres Rad 35°
Sturz	1°
Vorspur	4 - 6 mm

Bremsen
mechanisch durch Seile betätigte 2-Backenbremsen, Seile im Rahmen innen verlegt.

Bremstrommel	230 mm ∅
Bremsbacken	Breite 30 mm
Fußbremse	wirkt auf 4 Räder
Handbremse	desgleichen
Räder	4 Scheibenräder je 5 Bolzen
Felgen	3,00 D - 16
Reifen	Geländereifen 5,25 - 16
Aktiver Radius	319 mm
Radstand	2400 mm
Spurweite vorn	1356 mm
Spurweite hinten	1360 mm
Bodenfreiheit unbel.	310 mm
Bodenfreiheit belastet	290 mm
Rahmenhöhe unbelastet	330 mm
Rahmenhöhe belastet	310 mm

Gewichte

Fahrgestell
bestehend aus komplettem Rahmen, Motor, Getriebe, Hinter- und Vorderachse, Lenkung, Batterie mit Kabel und Anlasser = 485 kg

Militärisches Leergewicht
bestehend aus Fahrzeug, vollem Brennstoffbehälter, Öl, Ausrüstung = 790 kg

Typenleergewicht
bestehend aus militärischem Leergewicht abzüglich Ersatzrad und Ausrüstung = 745 kg

Militärisches Eigengewicht
bestehend aus Typenleergewicht abzüglich Seitenteile und Verdeck = 725 kg

Zulässiges Gesamtgewicht
bestehend aus militärischem Leergewicht plus zusätzlicher Belastung = 450 kg = 1240 kg

Ausrüstung
bestehend aus kompl. Werkzeug mit Tasche, 2 Paar Schneeketten, Kasten mit Ersatzteilen der elektr. Anlage, Wagenheber mit Kurbel, Andrehkurbel, 2 Montiereisen, Handlampe mit Kabel, Zeichenstab, Ölflasche = 28 kg

Achsdrücke aus dem zulässigen Gesamtgewicht

vorne	539
hinten	701
	1240

Achsdrücke aus dem militärischen Leergewicht

vorne	343
hinten	447
	790

Betriebsstoffe

Kraftstoff	OZ 74
Kraftstoffnormverbrauch	8,25 l/100 km
Kraftstoffbehälter	Fassungsvermögen 40 l einschl. 10 l Reserve
Fahrbereich	420 km
Lage	vorne im Aufbau
Zuleitung	Pumpe
Motoröl	Verbrauch 0,12 l/100 km Fassungsvermögen 3 l
Getriebeöl	Getriebefassungsvermögen 2½ l, Vorderachse-Fassungsvermögen 1½ l

Elektrische Anlage

Spannung: 6 Volt

Entstörung **einfach entstört**	**voll entstört**
Gruppe III	**Gruppe B I**
n. Bosch-Montageplan	n. Bosch-Montageplan
VTD 875	VTD 895 (Zündanlage)
	VTD 898 (Lichtmaschine)
	VTD 804 (Bremsl.Schalter)

Lichtmaschine	
»Bosch« RED $\frac{130}{6}$ 2600 A1 15 P (spannungsregelnd)	»Bosch« REDN $\frac{130}{6}$ 2600 A1 m. 1 Entstörer EM/S 75/1 u. 1 Entstörer EM/S 5/1
Verteiler	
»Bosch« VE 4 BS 276	»Bosch« VEN 4 (ZEZ) 7/93 B1 mit 1 Entstörer EM/S 5/1 u. Entstörgeschirr EM/GN 4
Zündspule	
»Bosch« TL 6/1	»Bosch« TKN 12/1 ohne Vorwiderstand
Zündkerzen	
»Bosch« 145 T 7	»Bosch« W 175 T 22
Scheibenwischer	
»S.W.F.«	»S.W.F.« – vollentstört,
2 Stk. (Schalter eingebaut) (nicht entstört)	
NSW/WH 6 Volt	6 V
Bremslichtschalter »Stots« nicht entstört	entstört mit »Bosch« Entstörer EMS 6/6/1
Batterie	6 V, 75 Ah Typ DIN Kr 2311
Anlasser	»Bosch« RED 0,4/6 L 3 (P)
Scheinwerfer	»W.M.I.«-Aufbauscheinwerfer 160 ∅ Lichtaustritt (Fernl./Abblendl./Standl.)
Tarnscheinwerfer	»Notek« (Stufenschalter hierzu auf der Schalttafel)
Heckbeleuchtung	»Notek«-Abstandrücklicht und »Notek«-Zusatzrücklicht
Lichtschalter und Zündschalter	in der Schalttafel (ebenso die Fernlicht-Kontrollampe)
Abblendung	»Kostal«-Fußabblendschalter
Winker	»S.W.F.«-Anbauwinker – 180 Pfeillänge
Winkerschalter	neben der Schalttafel (Winkerkontrollampe i. d. Schalttafel)
Warngerät	»W.M.I.«-Preßstoff-Tellerhorn 6 Volt
Sicherungen	10 Patronen 15/8 in 2 Kasten vorne 3 Patronen 15/8 in 1 Kasten hinten
Handleuchte	mit Steckbuchse a. d. Schalttafel
Lade-Kontrollampe	i. d. Schalttafel
Öldruck-Kontrollampe	i. d. Schalttafel (leuchtet mit eingeschalt. Zündung bei weniger als 0,8 atü Öldruck
Sonstiges	
Geschwindigkeitsmesser	»Veigel« bzw. »V.D.O.«, 100 ∅ Anzeigebereich 0 bis 100 km/st Antrieb über Schraubenräder und biegsame Welle vom vorderen Triebling

Schlepperfahrzeug Typ 82 mit Pak 37 (Typ 276)/1944

Veränderungen gegenüber dem Normal-Fahrzeug-Typ 82:

Motor	
Bohrung	75
Hub	64
Hubraum	1130 cm^3
mit Fliehkraft-Drehzahlbegrenzer an der Lichtmaschinenachse.	Ausführung lt. Stückliste, wie bei Typ155
max. Motordrehzahl	3300/min

Geändertes äußeres Stirnrad-Übersetzungsgetriebe an der Hinterachse

Übersetzung	14 : 29 = 1 : 2,07 V = 54 km bei n = 3300 Motordrehzahl V = 49 km bei n = 3000 Motordrehzahl
war früher	(15 : 21) = (1 : 1,4) (V = 80 km bei n = 3300 Motordrehzahl
Stirnrad-Achsabstand	a = 88,5 (war 76,5)
Mitte Hinterrad dadurch	12 mm höher verlegt

Räder und Reifen:	
Conti-Geländereifen	BF 200 x 16
effektiver Radius	R = 320 mm
Scheibenrad	82.68.01.2
Felge	7,00 . D x 16
Einpreßtiefe	13 mm (Normalrad 33)
Spurvergrößerung	+ 40 mm
Spurweite vorne	1396
Spurweite hinten	1400
1 kompl. bereiftes Rad wiegt	23 kg
Luftdruck vorne	1,2 atü
Luftdruck hinten	1,5 atü

Betriebsstoffverbrauch

Kraftstoffnormverbrauch	10 l/100 km
Ölverbrauch	0,15 l/100 km
Fahrbereich nach Norm Verbrauch	ca. 400 km

Fahrleistung in Ebene und Steigung
siehe Bericht vom 3. 11. 1944

Fahrgeschwindigkeit in den verschiedenen Gängen

	bei n = 3300	n = 3000 Motordrehzahl
1. Gang 1 : 3,6	12 km/h	10,9 km/h
2. Gang 1 : 2,07	20,8 km/h	19,0 km/h
3. Gang 1 : 1,25	34,6 km/h	31,4 km/h
4. Gang 1 : 0,80	54 km/h	49 km/h
Rücklauf 1 : 6,6	6,5 km/h	5,95 km/h

Gewichte

Fahrzeug-Leergewicht (milit. Marschgewicht)	760 kg
Fahrzeug zul. Ges.-Gewicht	1210 kg
Fahrzeug zul. Zuladung	450 kg
Fahrzeug Vorderachsbel. 40%	484 kg
Fahrzeug Hinterachsbel. 60%	726 kg

Zugkraft-Messung:

Zugkraft an der Anhängerkupplung, gemessen auf ebener Makadamstraße	Z = 700 kg beim Anfahren Z = 500 kg ca. 8 km/h

Geschlepptes Geschütz
Pak 37

Bereifung	6,00 - 20 Transport
Spurweite (gemessen)	1425 mm
Scheibenräder: Einpreßtiefe	27,5 mm
Bolzen-Zahl	5
Lochkreis	160 ∅
1 Rad wiegt bereift	43 kg
Ges.-Gewicht des Geschützes	565 kg

Geschleppter Munitions-Anhänger

Anhänger-Fahrgestell	F. Hahn-Zeichnung. 121.208. A
Anhänger mit Aufbau	276.00.010
Deichsellänge Mitte Öse bis Mitte Rad	1720 mm
Spurweite	1340 mm
Bereifung (wie bei Fahrzeug)	BF 200 x 16

Aufbau:
Holzpritsche innen mit 1 mm Blechbeschlag

lichte Breite		1025 mm
lichte Länge		810 mm
lichte Höhe		350 mm
Bodenfläche		0,82 m^2
Volumen		0,29m^3
Bodenfreiheit belastet	ca.	320 mm
Ganze Höhe	ca.	770 mm

Gewichte:

Anhänger-Fahrgestell mit Reifen	100 kg
Aufbau mit Kotflügel	50 kg
Ges.-Gewicht, leer	150 kg

Zuladung

16 Munitionskörbe je	393 mm lang) 125 mm breit) lt. O.K.H. 395 mm hoch)
Geschosse je Korb	3
Gewicht je Korb	25 kg
Anordnung der Ladung	stehend, 2 Reihen je 8 Körbe
in der Breite der Pritsche	8 x 125 mm = 1000 mm
in der Länge der Pritsche	2 x 393 mm = 786 mm
Gesamtgewicht der Ladung: (16 x 25)	400 kg
Gesamtgewicht des beladenen Anhängers	550 kg

Besatzung des Fahrzeuges

Geschützfahrzeug und Anhängerfahrzeug	7 Mann incl. Fahrer
Einsatzgewicht 1 Mann incl. Gepäck	100 kg

Außer den schon erwähnten VW Typen entstanden in den Jahren 1938-1945 auf der Basis der Karosserie der VW Kübelwagens und seines Motors eine ganze Reihe *»Sonder-Konstruktionen«:*

Typ Nr.:

87 VW — ein Geländewagen mit Allradantrieb auf der Basis VW 82. Beim Typ »87« handelte es sich um eine Weiterentwicklung des Typs »86«. Dieses offiziell als »VW-Geländewagen B« bezeichnete Fahrzeug hatte vier angetriebene Räder und den normalen viersitzigen, offenen Aufbau der Type »82«. Als Typ »87« hatte das Fahrzeug den geschlossenen vier-

sitzigen »KdF-Aufbau«. Für den Einsatz in Afrika wurde das Fahrzeug »tropenfest« gemacht. Die Umbauten erstreckten sich auf intensiven Staubschutz für Motor und E-Anlage. Eine Kraftstoffhilfspumpe mit Zusatzfilter war ebenfalls vorgesehen. Ferner wurde die Normalbereifung 5,25-16 gegen Sandreifen der Größe 22 mal 67 ausgetauscht. Das verlangte eine Umstellung der Vorderachse. Diese Fahrzeuge gelangten ausschließlich in Nord-Afrika zur Verwendung, jedoch wurden Restbestände vom Frühjahr 1943 ab auf den europäischen Kriegsschauplätzen eingesetzt.

92 VW	Pritschenwagen
110 VW	Kleinschlepper A, mit Führersitz-Anordnung (1938/39)
128 VW	Schwimmwagen A, groß
138 VW	Schwimmwagen B, groß
162 VW	Geländewagen, Typ selbsttragend
164 VW	Sechsrad-Gelände-LKW, Typ leicht mit 2 VW-Motoren, 1941 im Auftrag des SS-Führungsamtes entworfen
166 VW	Schwimmwagen C, klein, 1941 im Auftrag des SS-Führungshauptamtes, ein Krad-Wagen, kurzer Radstand aus Typ »87«.
174 VW	Ein Sturmboot mit VW-Motor
177 VW	5-Gang-Getriebe A für VW-Geländewagen
178 VW	5-Gang-Getriebe B für VW-Geländewagen (vereinfachte Ausfertigung)
179 VW	Motor mit Benzineinspritzung
182 VW	Geländewagen, 2-Radantrieb, mit Einheitsaufbau
187 VW	Geländewagen, Allradantrieb mit Einheitsaufbau
230 VW	mit Generator-Betrieb
231 VW	mit Gasgenerator (Acetylen)
235 VW	mit Elektroantrieb
239 VW	mit Holzkohlengenerator
240 VW	mit Flaschengas-Betrieb
276 VW	Typ »82« mit Protzhaken
287 VW	Kommandeurwagen
309 VW	Geländewagen, Allradantrieb, mit Einheitsaufbau
330 VW	mit HoKo-Gen.-Anlage
331 VW	mit HB-Gen.-Anlage
332 VW	mit AK-Gen.-Anlage
821 VW	3sitziger Kübelwagen-Aufbau
822 VW	Sirenenwagen-Aufbau
823 VW	Panzerkampf- und Panzerspähwagen-Aufbau
825 VW	Pritschenwagen für Fahrgestell Typ »82« und »92«
826 VW	Tropenkastenwagen
827 VW	Kommandeurwagen 3-sitzig, Fahrgestell Typ »82«
828 VW	offener Geländewagen aus Typ »820« in Holzausführung
870 VW	Geländewagen 4-sitzig, Typ selbsttragend
871 VW	Geländewagen 3-sitzig, Typ selbsttragend
877 VW	Kommandeurwagen auf Fahrgestell Typ »287«

Unzählige weiter Versuche und Entwicklungen hatten unmittelbar mit dem Volkswagen zu tun. So fand z. B. 1939/40 der VW-Motor der Type 120/121/122 als Stromerzeuger bei Heer, Luftwaffe und Reichspost Verwendung.

BIBLIOGRAPHIE:

Bittorf, Wilhelm: Die Geschichte eines Autos. In: Der Spiegel. Hamburg 1957

Cosson, L.: Ein Auto wird geboren. Porsches Leben. Antwerpen 1964

Delonge, Th.: Kraftfahrzeugkunde für den VW-Fachmann. Hannover 1966

Frankenberg, Richard v.: Ferdinand Porsche, der Weg eines genialen Konstrukteurs. Zürich–Leipzig–Wien 1957

German Army Transport, Part I.: Wheeled Vehicles. Automobile Engineer. London 1945

Glossary of Captured German; War Office, 1945 London

Gretland, Peer: Historien om Folkvognen Folkevognen. Oslo 1967

Hünninghaus, Kurt: Magier des Automobils – Ferdinand Porsche. München 1962

Index to German Military Instructional and Spare Parts: Handbooks for Wheeled and Tracked Vehicles, R. E. M. E. Army Division, Control Commission for Germany. London 1945

Kaes, Ghislaine E. J.: Die Nordamerikareise des Herrn Dr. Ing. h. c. Ferdinand Porsche im Jahre 1936. Vortrag vom 29. 1. 1937

Kaes, Ghislaine E. J.: Ferdinand Porsche, Mensch und Technik, Wirken und Schaffen. Vortrag vom 4. Oktober 1969

Kaes, Ghislaine E. J.: Ferdinand Porsche – Einhundert Jahre nach seiner Geburt. Stuttgart 1975

Karlmann Brechenmacher, Josef: Porsche, eine namensgeschichtliche Darstellung. Stuttgart 1939

Kölling, Fritz: Ein Auto zieht Kreise – Herkunft und Zukunft des Volkswagens. Reutlingen 1962

Meisel, Charles/Frankenberg, Richard v.: Porsche, the Man and his Cars. London 1961

Mors, Dr. Eberhard: Dein KdF-Wagen. Berlin 1938

Müller, Peter: Ferdinand Porsche ein Genie unserer Zeit. Graz – Stuttgart 1965

Nelson, Walter Henry: Die Volkswagenstory. München 1966

Oberkommando des Heeres, Heeres-Waffen-Amt: Leichter Personenkraftwagen, Baumuster K 1 Typ 82. Gerätebeschreibung und Bedienungsanweisung zum Fahrgestell. Berlin 1940

Oberkommando des Heeres, Heeres-Waffen-Amt: Leichter Personenkraftwagen K 1 Typ 82, Sonderausrüstung für Tropen. Berlin 1941

Oberkommando des Heeres, Heeres-Waffen-Amt: Leichter Personenkraftwagen K 1 Typ 82, Einbauanleitung zur Sonderausrüstung (Tropenwagen). Berlin 1941

Porsche, Ferry: Professor Porsche, 50 Jahre Arbeit für das Kraftfahrzeug. In: ATZ, Heft 6. Juni 1961

Post, Dan R.: Volkswagen Nine Lives Later. Arcadia, California 1966

Preston Charles: The Jokeswagen Book. New York 1966

Quint, Herbert A.: Porsche, der Weg eines Zeitalters. Stuttgart 1951

Reeres, Lt. Col. P. C.: The War-Time Activities of Dr. Ing. h. c. F. Porsche, Ministry of Supply R. T. R. British APO 413 Comlined Intelligence Objectives. Sub-Committee, G-2 Division, SHAEF, London'45

Reibestahl, Paul: Das Buch vom Volkswagen. Braunschweig 1951

Salus, Fritz Mitlo: Interview mit Professor Porsche. Dokumentation unserer Zeit. Salzburg 1969

Scholz, Hugo: Herr seiner Welt – der Lebensroman Ferdinand Porsches. Augsburg 1962

Sinz, Herbert: Der junge Porsche. Göttingen 1968

Sinz, Herbert: Autos – Seine Leidenschaft. Dortmund 1968

Sponsel, Heinz: Porsche – Autos – Weltrekorde. Bleckede 1952

Volkswagenwerk GmbH: Plaudereien um den KDF-Wagen und das Volkswagenwerk. Berlin 1939

Volkswagenwerk GmbH: VW Typ 82. Kurze Anleitung für den Werkstattmann über die wesentlichen Instandsetzungsarbeiten. Braunschweig 1940

Volkswagenwerk GmbH: Leichter Personenkraftwagen K 1 Typ 82, Ersatzteilliste. Berlin 1941

Volkswagenwerk GmbH: Leichter Personenkraftwagen K 1 Typ 82, Sonderausrüstung für Winterbetrieb. Braunschweig 1942

War Office: Technical Intelligence Summaries. Number 1 (February 1941) to Number 187 (November 1945). London 1945

War Office: Recognition of Technical Equipment. Used by the German Army (and other ground forces) Section III. Vheeled (unarmoured) Vehicles, SHAEF. London 1945

ZEITSCHRIFTEN:

Alte Kameraden (1955–1960)

Feldgrau. Mitteilungen einer Arbeitsgemeinschaft zur heereskundlichen Forschung. Burgdorf/Hannover 1953–1964

Der Frontsoldat erzählt. Flensburg 1953–1956

Heeres-Verordnungsblatt. Berlin 1932–1944

Wehrkunde. Monatsschrift der Gesellschaft für Wehrkunde. München 1953–1964

BILDQUELLEN

VW-Archiv (6)
Porsche-Archiv (15)
Bundesarchiv Koblenz (177)
Imperial War Museum (11)
Établissement Cinématographique et Photographique des Armées, Fort D'Ivry (1)
Archiv J. Piekalkiewicz (4)

EIN WORT DES DANKES

Ich möchte für ihre freundliche Hilfe meinen herzlichen Dank sagen:

Herrn Ghislaine E. J. Kaes, Stuttgart
Herrn Ing. H. Kaes, Salzburg
Herrn Dr. B. Wiersch, VW-Archiv, Wolfsburg
Herrn Dr. M. Haupt, Bundesarchiv Koblenz
allen Herren der Photographic Library, Imperial War Museum, London
Mr. J. S. Lucas und Mr. P. H. Reed, Imperial War Museum, London
Capitaine Brignone, Chef de la Division Photographique, Fort D'Ivry, Paris